PROFIT SHARING IN CANADA

PROFIT SHARING IN CANADA

The Complete Guide to Designing and Implementing Plans that Really Work

David E. Tyson

JOHN WILEY & SONS
Toronto • New York • Chichester • Brisbane • Singapore

Copyright © 1996 by David E. Tyson.

All rights reserved. No part of this work covered by the copyrights herein may be reproduced or used in any form or by any means — graphic, electronic or mechanical — without the prior written permission of the publisher.

Any request for photocopying, recording, taping or information storage and retrieval systems of any part of this book shall be directed in writing to the Canadian Reprography Collective, 6 Adelaide Street East, Suite 900, Toronto, Ontario, M5A 1H6.

Care has been taken to trace ownership of copyright material contained in this text. The publisher's will gladly receive any information that will enable them to rectify any reference or credit line in subsequent editions.

John Wiley & Sons Canada Limited
22 Worcester Road
Etobicoke, Ontario
M9W1L1

Canadian Cataloguing in Publication Data

Tyson, David
    Profit sharing in Canada

Includes bibliographical references and index.
ISBN 0-471-64146-4

1. Profit-sharing – Canada.   I. Title

HD2988.T98 1996 658.3'225 C95-933192-1

Quattro Pro is a registered trademark of Borland International. Microsoft Excell is a registered trademark of Microsoft Corporation. Lotus and 123 are registered trademarks of Lotus Development Corporation.

Production Credits
Cover & text design: Christine Rae
Printer: Webcom Ltd.

Printed in Canada
10 9 8 7 6 5 4 3 2 1

DEDICATION

This book is dedicated to the memory of my parents, Albert Edmund Tyson (1901-1973) and Vera O. Woods (1908-1989). They provided me with a love of reading and the written word that enabled me, with a little experience in the field, to write this book.

CONTENTS

Preface   xiii

Acknowledgements   xvii

Chapter 1: History and Prevalence of Profit Sharing   1
   Conclusion   5

Chapter 2: Types of Profit-Sharing Plans   7
   Definition of Profit Sharing   7
   Cash Profit-Sharing Plans   8
   Deferred Profit-Sharing Plans   8
   Employees Profit-Sharing Plans   9
   Combination Plans   9
   Advantages and Disadvantages   9
      Cash Plans   9
      Deferred Profit-Sharing Plans   10
      Employees Profit-Sharing Plans   12
      Combinations   12

Chapter 3: Effectiveness of Profit-Sharing Plans   13
   How Chief Executives Feel About Profit Sharing   14

Impact on Corporate Performance	16
Effects on the General Reputation of the Company	17
Trends	19
Conclusion	19

Chapter 4: Conditions Needed for Profit Sharing — 21
Employee Relations	22
Externally Competitive Base Salaries	23
Internal Equity	24
Profits	24
Organizational Structure and Job Design	25
Stability	26
Communications	26
Management Commitment	27
Lack of Proper Conditions	29
Conclusion	29

Chapter 5: General Principles of Designing a Plan — 33
Introduction	33
Choosing an Objective	34
Type of Plan	35
Employee Involvement	36
Announcement	38
Copy-Cat Plans	40
First Meeting of the Profit-Sharing Committee	40
Employer Contributions to the Profit-Sharing Plan	43
Definition of Profits	43
Type of Formula	43
Amount	44
Frequency and Timing of Payments	46
Compensation Policy	47

Chapter 6: Membership in the Plan — 49
Category of Employment	50
Length of Service	51
Employees Who Leave the Company	52
Employees on Variable Pay Plans	53
Unions	54
Other Considerations	55

Chapter 7: Allocation: Dividing the Funds — 57
Earnings	58
Job Levels	60

Seniority or Length of Service	61
Attendance	63
Employee Contributions	64
Merit Ratings	65
Equal Distribution	66
Combinations	67
Length of Service and Earnings	67
Length of Service and Job Level	68
Merit Ratings and Earnings	69
How to Choose an Allocation Formula	70
Objective(s) of the Plan	70
Ease of Communication	70
Acceptability	70

Chapter 8: Testing Your Formula — 73

Assumptions	75
General Rules	77
Test of Reasonableness	79
Testing	79
Communications	80
Option A1: Allocation = Base Earnings Only	82
Option B1: Allocation = Job Levels/Equal Within Levels	83
Option C1: Allocation = Seniority/Length of Service	84
Option D1: Allocation = Attendance (Regular Hours Only)	85
Option E1: Allocation = Employee Contributions to Group RRSP	86
Option F1: Allocation = Merits Ratings	87
Option G1: Allocation = Equal Distribution	88
Option H1: Allocation = Length of Service/Base Earnings	89
Option I1: Allocation = Length of Service/Job Level	90
Option J1: Allocation = Merit Ratings/Base Earnings	91

Chapter 9: Deferred Profit-Sharing Plans — 93

Membership	94
Vesting	94
Forfeitures	95
Withdrawals	96
Investment Policy	98
Income Tax Act	99
Where to Invest	99
Employee Choice of Investments	100
Administration of Investments	101
Leaving the Plan	102
Contributions	103

Excess Contributions	103
Setting Up a Deferred Profit-Sharing Plan	104

Chapter 10: Employee Profit-Sharing Plans — 105

Membership	106
Vesting	106
Forfeitures	108
Withdrawals	109
Investments	109
Investment Choices for the Employee	110
Employer Contributions	110
Employee Contributions	111
On Leaving the Plan	111
Setting Up an Employees Profit-Sharing Plan	111

Chapter 11: Communications — 113

Communications During Design of the Plan	113
Communications After Implementation	114
Recruiting	114
Orientation	115
On Joining the Plan	115
Supervisory Training	116
Employee Meetings	116
Cheques	116
Pre-Retirement Planning	117
What to Communicate	118
Communications Media	120

Chapter 12: Administration — 123

Should There be a Committee?	124
Composition and Size of the Committee	124
Selection of the Committee	126
Term of Office of Committee Members	126
Transition from Design to Permanent Committee	128
Responsibilities of the Profit-Sharing Committee	128
Administration of the Plan	129
Reviewing the Plan	129

Chapter 13: The Design Process — 131

Introduction	131
Research and Planning	133
First Meeting of the Profit-Sharing Committee	135
Committee Members Meeting(s) with Employees	138

Second Meeting of the Profit-Sharing Committee	141
Developing and Testing of Models	144
Third Meeting of the Profit-Sharing Committee	145
Implementation	146

Chapter 14: Training of Employees — 147

General Training (All Employees)	147
Mechanics — How the Plan Works	147
Finance	148
Investments	149
Creative Thinking	150
Language Training	150
Employee Suggestion Systems	150
Interpersonal Skills	151
Performance Measures	151
Managerial/Supervisory Employees	152
Administration of Base Pay	152
Coaching Employees	153
Teams	153
Performance Management	153
Managing Diversity	154
Other Considerations	154

Chapter 15: Employee Suggestion Systems — 155

Why Have an Employee Suggestion System?	155
When to Establish an Employee Suggestion System	157
Designing a Suggestion System	158
Membership Eligibility	159
Process	160
Who Does the Evaluations?	162
Nature and Size of Awards	164
Protection of Suggestions	165
Changes in Equipment or Processes	166

Chapter 16: Pay-For-Performance and Profit Sharing — 167

Theories of Motivation	168
Expectancy Theory	168
Equity Theory	170
Definition of Pay-For-Performance	170
Types of Pay-For-Performance Plans	171
Productivity Gain-Sharing Plans	172
Differences Between Profit Sharing and Productivity Gain Sharing	173

Comparison of Types of Pay-For-Performance	174
Conclusion	175

Chapter 17: Profit Sharing and Unions/Organized Labour — 177
Introduction — 177
Canadian Labour Law — 177
Views of the Labour Movement — 178
Companies With Unions and Profit-Sharing Plans — 179
Installing a Plan in a Union Situation — 181
Trends in Union Shops and Profit Sharing — 184

Chapter 18: Profit Sharing in the United States and Mexico — 187
Introduction — 187
The United States — 188
Mexico — 189
Separate or Universal Plans — 191
Conclusion — 193

Chapter 19: Sample Profit-Sharing Plans — 195

Appendix 1: Additional Sources of Information — 213

Appendix 2: Formulas for Chapter 8 — 217

Index — 229

PREFACE

Profit sharing is an idea that has been around for more than 150 years and it has been described as old wine in new bottles. Profit sharing is defined as "a compensation program which makes payments to employees, over and above their base salaries or wages, which are determined by the level of profits of the corporation."

For example, Atlas-Graham Industries Co. of Winnipeg, Manitoba, a small manufacturer of brushes and mops, has a classic, cash, profit-sharing plan. The company deducts 2 per cent of sales from net profit before taxes and distributes 30 per cent of the remainder equally among the employees.[1]

Profit sharing is now more relevant than ever for Canadian companies. They are faced with the effect of free trade agreements, globalization, declining or stagnant productivity, shortages of skilled employees, and an increasingly demanding workforce seeking involvement and empowerment. Profit-sharing plans can assist companies to compensate employees without increasing fixed costs and at the same time increasing profitability and productivity.

---

[1] For a more complete description of this plan see Chapter 19, "Sample Profit-Sharing Plans."

This book is written for those employers and/or employees who are considering the introduction of a broad-based employee profit-sharing plan in their organization. Broad-based plans are those that include most employees and must be distinguished from "top hat" plans which are usually restricted to a few senior executives. It is intended to be a practical guide to the design and implementation of a profit-sharing plan that will enable the reader to install one with little or no help from external advisers.

My fundamental premise in writing this book is that profit-sharing plans are a form of compensation; using the term in its broadest application to include benefits, perquisites, salaries, wages, and even intangibles such as titles, parking spaces, and recognition programs.

Most of the literature on profit sharing in Canada emphasizes the payments, cumulative investments, and tax advantages of profit-sharing plans. This book concentrates on the motivational power of employee involvement that results when profit-sharing plans are properly designed, implemented and administered as part of a comprehensive, progressive, human resource strategy. The tax implications are addressed in sufficient detail so that the reader is able to make any major decisions that depend upon the income tax consequences.

It is important to recognize that the definition provided earlier differentiates profit-sharing plans from "productivity gain-sharing plans" (the classic types are the Scanlon, Rucker and Improshare plans). This distinction is maintained throughout this book but it is not universal. Much of the literature on the subject of group incentive plans (which include both profit sharing and gain sharing) classify profit sharing as a form of gain sharing. I disagree with this categorization and treat both types of plans as forms of group incentive plans. This subject is discussed in more detail in Chapter 16, "Pay-for-Performance and Profit Sharing."

I want to make one final point. Throughout the book I frequently refer to "what most companies do"; or, "the most common approach," or even what I feel is the best way. However, you must remember that there is no absolute "right" way to design a profit-sharing plan. All design decisions must be made in the context of your objective(s); your corporate culture; the expectations, characteristics, and values of your employees; and, finally, the productivity and profitability of your firm. It is your company, your profit-sharing plan and you must fit the general principles of plan design to your circumstances. I hope that this book will help you do that.

ACKNOWLEDGEMENTS

Although I had been involved in profit sharing prior to starting my own management consulting business in 1983, I did not become active in the field until 1984. I was invited by Herb Brown, president of the Profit-Sharing Council of Canada, to join the council's roster of speakers for their seminar on Designing a Profit-Sharing Plan. Herb utilized my services in that regard until he retired in 1990. He had been president of the council since 1972 and did an incredible amount towards advancing the cause of broad-based employee profit-sharing plans in Canada. Herb was succeeded as president by Neil Walker, who is now president of Baxter Consulting. Neil continued my association with the council as a regular speaker. I am extremely grateful to both Herb and Neil for their support.

I must also acknowledge the contribution made by my fellow speakers at the council's seminars. They were: Duncan Macgregor (Peter Macgregor Meat & Seafood), Alex King (ERS Consulting), Mike Welch (Allstate Insurance) and Arch Brown (Canadian Tire Associate store, Barrie).

Bob Totten, of Totten & Associates Inc., in London, Ontario, a long time associate and friend, did me the courtesy of reviewing the manuscript for this book and offered many useful suggestions and ideas.

John Prior, president of J.P. Consultancy Inc., of Scarborough, reviewed several chapters, specifically the one on Employee Training.

Also, thanks to Bruce Pearce and the rest of the staff of the Newman Industrial Relations Library at the University of Toronto for helping me with my research.

Poncho Rodriguez, my friend and associate from Juarez, Mexico, did almost all my research on profit sharing in Mexico. His exceptional fluency in both Spanish and English was invaluable in obtaining and understanding the material.

Finally, I want to thank my clients. Without them neither my consulting practice nor this book would be possible.

1
HISTORY AND PREVALENCE OF PROFIT SHARING

The literature is somewhat divided on who devised the first profit-sharing plan. Some writers credit Jean Leclaire,[1] who owned and operated a house-painting business in Paris, France, and introduced a profit-sharing plan in 1842. Although the exact details are unclear, Leclaire's workers were very sceptical about the plan until he distributed 12,266 gold francs among them the following year. Apparently this business is still in operation today. Other writers have chosen Albert Gallatin as the inventor of the first profit-sharing plan in the United States, at his glass works in New Geneva, Pennsylvania, in 1797.[2]

Even large, well-known companies started profit sharing in the late nineteenth century; Procter and Gamble introduced its profit-sharing plan in 1887. Two other famous plans that still exist today are those of Eastman Kodak and Sears Roebuck, which were established in 1912 and 1916 respectively. It is also noteworthy that the Canadian subsidiaries of these companies also have profit sharing.

[1] Humfrey Michell, *Profit-sharing and Producers' Co-operation in Canada* (Kingston, Ont.: Jackson Press, 1918), 3.

[2] Charles R. Perry and Delwyn H. Kegley, *Employee Financial Participation: An International Survey* (Philadelphia: Industrial Research Unit, the Wharton School, University of Pennsylvania; 9104-6358; 1990), 129.

In Great Britain, one of the earliest plans was started at Briggs Collieries in 1865.[3] It paid out half of the profits in excess of a 10 per cent return on capital. Other early profit-sharing plans included those at South Metropolitan Gas Company (1889); Lever Brothers (1909); William Thomson and Sons of Huddersfield (1886); and Taylors (1892).

British interest in the subject was quite high and the Labour Department conducted annual studies of profit-sharing plans from 1891 to 1938, which were published in the *Labour Gazette*. For the purpose of these studies, the Labour Department used a definition of profit sharing that had been developed at the International Congress on Profit Sharing held in Paris in 1889. This definition specified that plans must include a large proportion of employees (not only supervisory staff or managers) and must pay them a bonus related to profits in addition to their wages. The formula for determining the amount of profits to be shared had to be predetermined. This definition is remarkably similar to my definition in this book.

One of the most interesting findings of the studies done by the Labour Department was the consistency of the size of the profit-sharing payments as a percentage of wages. The mean bonus between 1891 and 1938 was 5.1 per cent of wages; bonuses were consistently between 4 and 7 per cent.[4] Similar studies in 1954 and 1980 reported that the bonuses were 6.3 per cent and 5.1 per cent respectively. These numbers are also consistent with those recommended for use when determining the size of the profit-sharing pool in Chapter 6.

In 1988, Bell and Hanson estimated that in Great Britain 800 to 900 companies had profit-sharing plans, with about 2,250,000 participants.[5]

In Canada, the earliest profit-sharing plan that I could locate is that of The W.F. Hatheway Company of Saint John, New Brunswick, which was established in 1888 and operated until 1913 when it was "temporarily" suspended.[6] Employees with salaries of $1,000 to $1500 (which I assume are annualized

---

[3] Derek Matthews, "The British Experience of Profit-Sharing," *Economic History Review*, 2d ser., XLII, 4(1989), 440.
[4] Matthews, 449–450.
[5] D. Wallace Bell and Charles G. Hanson, *Profit Sharing and Profitability: How Profit Sharing Promotes Business Success*, (London: Kogan Page, 1989), 18.
[6] Michell, 8–9.

CHAPTER 1 — HISTORY AND PREVALENCE OF PROFIT SHARING

amounts) received 2 per cent of the company's net profits. Staff with salaries from $500 to $1,000 received 1 per cent while those "in the tea room and warehouse" were given 0.5 per cent to 1 per cent of the profits, depending on their wages.

Employees could choose to take the payment in cash or to leave it on deposit with the company, which paid interest at a rate of 6 per cent. Employees were only supposed to withdraw the monies for serious emergencies, such as buying a house, or on leaving the company.

Other early, Canadian profit-sharing plans reported by Michell include the following:

COMPANY	LOCATION	PLAN DATES (START/FINISH)	
John Morrow Machine Screw Company	Ingersoll, Ont.	1903–1908	
British Columbia Electric Railway Company	Victoria, Vancouver, New Westminster, B.C.	1902–1910	
Wortman & Ward Manufacturing Company	London, Ontario	? –1913	
Messrs. Stanley Mills & Co., Ltd.	Hamilton, Ontario	1903–	?
W.J. Gage Company	Toronto, Ontario		?
James Walker Hardware Company	Montreal, Quebec	1907–	?
The Steel Company of Canada	Hamilton, Ontario	1913–	?

As this list shows, profit sharing in Canada has been around for a very long time—over 100 years. The more interesting historical information is that profit sharing has shown spectacular growth in recent years. According to the 1992 Hewitt Survey, the distribution of plan start-up dates for 130 companies was as follows:[7]

DATE OF PLAN START-UP	PERCENTAGE OF RESPONDENTS
prior to 1950	2
1950–1959	2
1960–1969	6
1970–1979	20
1980–1989	48
1990–present	22

A survey by Richard Long of the University of Saskatchewan also reported similar growth in profit-sharing plans. Long's report showed that 58 per cent of the companies with plans had implemented them since 1980,[8] while 35 per cent had been initiated between 1986 and 1990.

What type and size of companies install profit-sharing plans? Both of the above surveys provide interesting, and similar, data on this subject. Hewitt reports the following distribution of plans by size of company in terms of the number of employees.[9]

NUMBER OF EMPLOYEES	PERCENTAGE OF COMPANIES WITH PLANS
1000+	36
500–999	10
100–499	30
under 100	24

The Long report provided data on 626 companies surveyed, of which 14.9 per cent had profit-sharing plans. Of the 626 companies, 83.4 per cent had 500 or fewer employees; the median size was 100 employees. The rate of unionization among the companies was 32.4 per cent, and 83 per cent were private corporations.[10]

---

[7] Hewitt Associates, *Survey of Canadian Profit-Sharing Plans,* (North York, Ontario, 1992), 8.
[8] Richard J. Long, *Employee Profit Sharing and Share Ownership in Canada—Results of a Survey of Chief Executive Officers* (Saskatoon, Sask. 1991), 15.
[9] Hewitt Survey, 1992, 8.
[10] Long, 2.

CHAPTER 1 — HISTORY AND PREVALENCE OF PROFIT SHARING

It is very difficult to determine how many Canadian companies have broad-based employee profit-sharing plans since there is no central voluntary or mandatory registry. However, both the Hewitt and Long surveys provide some indication of the extent of profit sharing in Canadian industry.

As stated above, the Long report showed that 14.9 per cent of companies had profit-sharing plans (all types), while 2.4 per cent had both profit sharing and an employee share ownership plan.[11]

The Board of Trade of Metropolitan Toronto also conducts a number of salary surveys, by occupational group, of companies in the Greater Toronto Area. One such survey, the Clerical Survey, reported that of 494 companies, 35 per cent had profit-sharing plans. The Information Technology Survey, however, surveyed 274 companies and found that 25.5 per cent had profit-sharing plans. These plans are probably all broad-based plans since such plans, by definition, include the type of employees in the survey.[12]

CONCLUSION

Three important points emerge from the above summary of the history of profit sharing. First, profit sharing is not a new idea. It originated over 150 years ago in France and has been in Canada for over 100 years.

Second, despite its ancient lineage, profit sharing seems to have received a new lease on life and is rapidly becoming a preferred form of compensation for the new, leaner companies of the 1990s.

Finally, early profit-sharing plans had as many variations and combinations as plans that are introduced today. Both the old and the new plans vary widely in terms of their objectives, membership rules, allocation formulas and types of payments. This means that profit-sharing plans can fulfil a wide variety of needs because of their flexibility, and still be effective.

---

[11] Long, 11.
[12] Metropolitan Toronto Board of Trade, *Clerical and Information Technology Salary Surveys* (1994), 47, 53.

2
TYPES OF PROFIT-SHARING PLANS

DEFINITION OF PROFIT SHARING

Before we continue further in our discussion, it is important to distinguish between "broad-based" and "top-hat" profit-sharing plans. This book addresses broad-based plans, although many of the principles can also be applied to designing and implementing top-hat plans.

Broad-based profit-sharing plans include most of the employees of the unit or organization for which the plan is designed. Top-hat plans, however, typically are reserved for a small group of senior executives. There are four major types of broad-based profit-sharing plans:

- cash
- deferred
- employees profit-sharing plans
- combination plans

All types of plans share three common features:
1. they establish a pool of money (the bonus pool or fund), which is calculated as a portion of profits of the unit or organization;
2. this pool of money is divided among the employees on the basis of a formula that is usually completely mechanistic; and
3. all company employees (with some exceptions—see Chapter 6) are involved, not just single departments or groups of employees, as is the case with productivity gain-sharing plans or top-hat profit-sharing plans.

CASH PROFIT-SHARING PLANS

In cash profit-sharing plans, the payouts are treated like any other cash payment to an employee and the company deducts personal tax at source. The payments are deductible to the employer as an expense just like salaries and wages. The *Income Tax Act* sets no limits on the size of the payments that the employee can receive or the employer can deduct.

DEFERRED PROFIT-SHARING PLANS

Deferred profit-sharing plans (DPSPs) are governed by section 147 and other sections of the *Income Tax Act*. Employers may make payments to a trust fund on behalf of employees up to certain maximums set by the Act. The Act limits both

(a) the total contributions made by an employer (on behalf of an employee) and the employee to DPSPs, Registered Pension Plans (RPPs), and Registered Retirement Savings Plans (RRSPs), and

(b) the total contributions made by the employer to a DPSP.

Within these limits, the employer can treat the payments as deductible expenses. The employer is not required to withhold personal income tax nor is the employee required to pay it. The trust fund can be invested in most types of investments although there are some limits, especially for Canadian-controlled private corporations (see Chapter 10). Earnings and capital gains of the fund are not taxed until they are withdrawn—exactly like RRSPs.

EMPLOYEES PROFIT-SHARING PLANS

An employees profit-sharing plan (EPSP) is established under section 144 of the *Income Tax Act*. The term "employee" is misleading in this instance since all plans described in this book include employees. Employees profit-sharing plans are similar to DPSPs in that monies are placed in a trust, by the employer, on behalf of the employee. The major difference between the plans is that there are no limits on the amount that the employer can contribute to the plan. However, personal income tax must be paid (i.e., withheld) by the employer. There are no limits on the types of allowable investments, and the earnings and capital gains of the fund are taxed each year. Furthermore, companies are not legally required to register the plan with Revenue Canada, as they are with DPSPs.

COMBINATION PLANS

Employers may also choose to create a profit-sharing plan that is a combination of the three main types discussed above. For example, many companies combine a DPSP with an EPSP so that, when the allowable contribution limits for the deferred plan are reached, the excess is placed into the EPSP. Another common option is to pay a certain percentage (e.g., 50 per cent) in cash and put the remainder into a deferred plan. This allows the employee to have some cash to meet immediate needs but also to put away some funds for the future. The major features of any combination plan will depend on the exact mixture of the three major types chosen.

ADVANTAGES AND DISADVANTAGES

There are no clear-cut choices in selecting the type of profit-sharing plans to be implemented. Each of the four major options offers various advantages and disadvantages.

CASH PLANS

Cash plans are typically considered to have more direct motivational impact on employees since the payout occurs relatively close in time to the performance involved. It is a common principle of behaviour modification that if an employer wants to encourage performance of a certain behaviour, then that behaviour should be rewarded as soon as possible after it has occurred. Another advantage of cash plans is that most employees will choose one if they are given the option. For this reason, they are more acceptable and easier to sell initially.

A disadvantage of cash plans is that the motivational impact of the cash payment does not last long. In companies where the payment tends to be very regular, and/or paid at Christmas, employees often view it as a Christmas bonus and the connection between the performance of the organization (and its employees) is lost. The connection is also blurred when the payment is made three or four months after the end of the fiscal year. This could mean that actions by the employee at the start of the fiscal period are separated by as much as 15 to 16 months from the actual cash payment. Allocations under a cash plan are taxable and the employee only sees the after-tax portion of his or her share of the profits.

Many companies start their profit-sharing program with a cash plan but later find that a deferred plan (usually a DPSP or a combination cash/DPSP) is better suited to their needs. However, it can be difficult to make this transformation without increasing the proportion of profits put into the pool. There are two reasons for this. First, the existing cash plan may be using up all the profits that you are prepared to contribute. To avoid decreasing the cash income of employees, you must stack the deferred plan on top of the cash plan, which simply requires more money.

Second, the payouts to the employees from the cash plan may be just sufficiently large that employees consider them to be motivational. In both cases the only solution is to increase the company contribution to the profit-sharing pool.

Deferred Profit-Sharing Plans

Deferred profit-sharing plans (DPSPs) have the great advantage that the employer contributions to the fund are not taxable to the employee (up to the allowable limits), yet are still a deductible expense for the employer. An even greater advantage is that the earnings and capital gains of the trust are not taxable until they are removed from the plan. The relatively greater increase in growth of tax-sheltered funds as opposed to taxed funds is well known. The longer the monies are in the tax shelter, the greater this difference becomes; it is virtually exponential.

A seldom-mentioned advantage of DPSPs is that, if properly invested, they often will increase in value even when the company is not profitable and a cash plan would make no payouts at all. For example, suppose the typical employee at a company has an average vested amount in the DPSP of $10,000. Under "normal" circumstances, the employee might receive $250, which is placed in the trust and the $10,000 would generate $1,000 in earnings (assuming that 10 per cent bonds are available). If the company does not make any profits, the $250 will not materialize but the $1,000 will accrue.

A further advantage is the fact that DPSPs can serve as retirement vehicles (either instead of, or in addition to, a registered pension plan) with fewer regulations and fixed obligations. They can also be a form of long-term disability insurance since employees may make withdrawals from DPSPs, although they must pay income tax on the amounts withdrawn. This may not be as onerous as it sounds because they can withdraw their funds in stages and are probably, because of their circumstances, in a lower tax bracket anyway.

A disadvantage of DPSPs is that many employees, especially younger ones, do not consider them attractive—at least initially. A 25-year-old employee is as interested in a deferred plan as he or she is in a pension plan or life insurance. These plans are more difficult to sell initially and generally take longer to become accepted and understood. They are also more complex and require more administrative time and effort.

It is important to recognize, however, that in spite of what may appear to be significant disadvantages, many companies in Canada make at least part of their profit-sharing plan a DPSP and many convert to such a plan after starting with a strictly cash plan.

A further disadvantage of DPSPs is that the *Income Tax Act* places restrictions on the investments that may be made under such a plan (see Chapter 9). Allowable investments are listed in section 204(e) of the Act. The most limiting restriction prevents DPSPs installed by a Canadian-controlled private corporation from investing in their own stock and from having "significant" shareholders as members of the plan at all.

Employees Profit-Sharing Plans

Employees profit-sharing plans have an advantage over DPSPs in that there are no limits on the investments of the fund. Any investment vehicle is permissible, including foreign securities, even if the employer is a Canadian-controlled private corporation. Furthermore, EPSPs do not have to be registered with Revenue Canada, as DPSPs do.

The major disadvantage of EPSPs is that the monies allocated to members of the plan are taxed at the time of allocation and the earnings and capital gains are also taxed as they occur. This means that the growth of the fund is dramatically slower than for a DPSP. Nevertheless, some Canadian-controlled private corporations have found that, if they want employees to share directly in the business's success by owning shares in the employer company, an EPSP is the right choice. These plans can also invest in the stock of a private corporation, whereas DPSPs cannot.

Combinations

Combination plans share all the advantages and disadvantages of the subplans that comprise them. The main feature of combination plans is that they can meet a wider variety of objectives and thus satisfy a greater number of employees. On the other hand, combination plans are more complex and therefore more difficult to administer and communicate. They also usually require higher employer contributions to the fund than a single type of plan since it requires more money to achieve more objectives.

This aspect of plan design regarding how much money is required from the employer is discussed in more detail in Chapter 4.

3
EFFECTIVENESS OF PROFIT-SHARING PLANS

Anyone who is considering the introduction of an employees profit-sharing plan always wonders what effect it will have on the organization. In other words, to what extent will the company be more profitable or productive if such a plan is developed? I believe that the most valuable insight on this subject comes from Robert Shad, founder and president of Husky Injection Moulding Systems Ltd. of Bolton, Ontario, who has personal experience with profit sharing. He explains, "Profit sharing doesn't cost a thing as far as I am concerned. The company just makes more money if you have such a plan—because you get better people and better performance."[1]

In this chapter we will look at surveys of the opinions of other chief executives and human resources officials in companies with profit-sharing plans. We will also examine the results of a major study comparing the performance of companies with and without profit-sharing plans. We will also consider another method to assess the effectiveness of profit sharing, which is to

[1] Eva Innes and others, *The 100 Best Companies to Work for in Canada* (Toronto: Harper Collins, 1986), 349.

look at the general reputation of companies with plans—are they considered good employers?

HOW CHIEF EXECUTIVES FEEL ABOUT PROFIT SHARING

Richard Long, whose work was discussed in Chapter 1, contacted the chief executive officers of 626 Canadian firms between May 1989 and June 1990[2] to gather the following information: whether their company had a broad-based profit-sharing plan or an employee share ownership plan; and the effectiveness of these plans based on a number of criteria. The CEOs were asked to rate the profit-sharing plan using the following scale, with -5 as extremely negative to +5 as extremely positive. A rating of 0 meant the plan had no impact on that factor. The mean (average) results are shown below:[3]

FACTOR IMPACTED	PERCEIVED IMPACT OF PROFIT SHARING − MEAN SCORE
The company overall	+3.63
Employee job satisfaction	+3.02
Employee loyalty	+3.25
Employee motivation and effort	+3.43
Employee interest in firm performance	+3.62
Ability to recruit employees	+2.08
Employee turnover	+2.64
Employee absenteeism	+1.75
Cooperation within firm	+2.56
Industrial relations	+1.91
Grievance rates	+1.38
Company profitability	+2.58
Value of company stock	+1.23

None of the CEOs in this survey reported a negative impact on any of the above factors. These are impressive results.

Additional, similar data are available on the incidence and effectiveness of profit sharing in the United States. One of the

---

[2] Richard J. Long, "The Incidence and Nature of Employee Profit Sharing and Share Ownership in Canada," *Industrial Relations/Relations Industrielles* 47, no.3 (1992): 471.
[3] Long, 22.

more interesting studies is *People, Performance and Pay*,[4] which reviews a number of reward and other human resource practices that are described as "non-traditional." The report studies such rewards as profit sharing; gain sharing; lump-sum bonuses; individual incentives; all salaried workforces; pay for knowledge (also known as skill-based pay); small group incentives; and earned time off in 1598 companies.[5]

The data in this report must be considered carefully since they do not separate broad-based plans from top-hat ones, as does the Long study. This report includes cash profit sharing at the unit level (e.g., division) in the definition of gain sharing and defines profit sharing as an "annual bonus or share based on company/corporate performance [that] could be paid in cash or deferred into a retirement fund."[6] Nevertheless, it is possible to differentiate broad-based profit-sharing plans in this report.

Two hundred and twenty-three organizational units reported that they had a gain-sharing plan, of which 21 per cent were cash profit sharing.[7]

The reasons for introducing gain sharing (which includes profit sharing) are very interesting and include the following:[8]

- productivity improvement
- quality improvement
- better employee relations
- labour cost reduction
- pay for performance
- competitive pressures
- HR philosophy/culture shift
- union avoidance
- corporate mandate
- bargaining trade-off

---

[4] Carla O'Dell and Jerry McAdams, *People, Performance and Pay: A Full Report on the American Productivity Center/American Compensations Association National Survey of Non-Traditional Reward and Human Resource Practices*, (Houston: American Productivity Center, 1987).
[5] O'Dell and McAdams, 8.
[6] Ibid, 35.
[7] Ibid, 34.
[8] Ibid, 37.

16 PROFIT SHARING IN CANADA

People, Performance and Pay also provides some data on the performance of the profit-sharing plans in its definition of gain sharing. The authors asked the participating companies to rate their plan in terms of the impact of the plan on a number of factors, using a scale.[9] The results are shown in the following table.

IMPACT ON	PERCENT REPORTING POSITIVE (4) OR VERY POSITIVE (5)*
Productivity	65
Costs	56
Quality	70
Scrap and rework	39
Labour relations	73
Employee involvement	70
Communications	77
Employee's pay	77
Work rules	27
Turnover	56

* The scale was 1=very negative impact; 2=negative; 3=no impact; 4=positive; 5=very positive impact.

IMPACT ON CORPORATE PERFORMANCE

In Great Britain, D. Wallace Bell and Charles G. Hanson[10] conducted a study that was published in 1987, and revised in 1989. They studied 414 British companies and compared the performance of those with profit sharing (113) with those that did not have such a program (301). The results are as follows:[11]

PERFORMANCE FACTOR*	PROFIT-SHARING COMPANIES	NON-PROFIT-SHARING COMPANIES	PERCENTAGE DIFFERENCE
Return on equity (%)	25.1	19.9	26.1
Return on capital employed (%)	20.6	15.5	32.9
Earnings per share (pence)	16.3	12.8	27.3

[9] O'Dell and McAdams, 38.
[10] D. Wallace Bell and Charles G. Hanson, *Profit Sharing and Profitability: How Profit Sharing Promotes Business,* (London: Kogan Page, 1989).
[11] Bell and Hanson, 58.

PERFORMANCE FACTOR*	PROFIT-SHARING COMPANIES	NON-PROFIT-SHARING COMPANIES	PERCENTAGE DIFFERENCE
Return on sales (%)	8.4	5.6	50.0
Annual growth in sales (%)	15.5	13.7	13.1
Annual growth in equity (%)	17.6	16.0	10.0
Annual growth in profit (%)	13.6	9.7	40.2
Dividends per share (pence)	5.2	4.9	6.1
Total annual returns (%)	24.8	18.0	37.8

* These are the average ratios for the eight years from 1977/78 to 1984/85.

Further evidence of the effectiveness of profit-sharing plans is provided in the paper "Paying for Productivity: A Look at the Evidence,"[12] in which the editor, Alan Blinder, reports on a number of studies on profit sharing (including Bell and Hanson). Five of the six studies include American companies in a variety of industries and were conducted between 1971 and 1979. All of these studies show that profit-sharing companies perform better, on as many as 144 points of comparison, than companies without profit sharing.[13] Blinder's comment on one study sums up all the others: "Profit-sharing companies do better than the others."

EFFECTS ON THE GENERAL REPUTATION OF THE COMPANY

Another method we can use to judge the effectiveness of profit-sharing plans is the general reputation of companies with profit-sharing plans. This is not to say that the credit for a good reputation can be attributed solely to a profit-sharing plan, but those companies that are considered to be good employers often have such plans. The high incidence of profit sharing among such employers is quite striking.

[12] Alan S. Blinder, ed., "Paying for Productivity: A Look at the Evidence," (Washington: The Brooking Institute, 1990).
[13] Blinder, 125.

For example, consider the two books, *The 100 Best Companies to Work For in Canada* and *The 100 Best Companies to Work For in America*, in which the authors set out to identify the best employers in each country.[14] In the Canadian book they identified a number of characteristics (based on interviews with employees) that tended to make the companies good employers. These qualities included the following:[15]

- team spirit—a high level of co-operation and regard for the individual;
- communications—a variety of programs that go from top to bottom and vice versa;
- incentives—including *profit sharing*,[16] gain sharing, and stock ownership plans;
- above-average pay and benefits; and
- good union relations.

The similarity between the above list and the conditions required before installing a profit-sharing plan[17] is impressive.

Of the 100 best companies in Canada, 43 have some form of broad-based employees profit-sharing plan; 15 have a stock ownership plan; and 57 have some form of group incentive plan. Many companies—such as Apple Canada, DMR, Magna International, Procter and Gamble Inc., Dofasco Inc., London Life Insurance Co., Four Seasons Hotels, S.C. Johnson Wax, and Kodak Canada—are also leaders in their respective fields.

The 100 Best Companies to Work For in America, which was the model for the Canadian edition, uses a similar methodology and came up with similar results. Forty-nine of the companies listed in the 1993 edition have a broad-based profit-sharing plan, 16 have a stock ownership program, and three have a gain-sharing system. Nine companies have at least two of the three types of incentives listed by the authors.

---

[14] For more information see Eva Innes et al, *The 100 Best Companies to Work for in Canada* (Toronto: Harper Collins, 1986) and Robert Levering, *The 100 Best Companies to Work for in America* (New York: Penguin Books, 1994).
[15] Innes et al., 6-9.
[16] My italics.
[17] See Chapter 4, "Conditions Needed for Profit Sharing."

TRENDS

Another way to judge the effectiveness of profit-sharing plans is to look at trends regarding the numbers of plans that have been introduced and the intentions of executives regarding the development of new profit-sharing plans. Richard Long's survey (referred to in Chapter 1) published in 1991, reported that 58 per cent of the profit-sharing plans had been introduced since 1980; 35 per cent were installed from 1986 to 1990.[18] Long also reported that 56 of 513[19] companies (10.9 per cent) indicated that they intended to introduce profit sharing within the next two years. The Hewitt Survey in 1992 reported similar data.[20]

CONCLUSION

This chapter has examined evidence from Canada, the United States, and Great Britain—three countries with similar business cultures and economic and political traditions. We have looked at the opinions of chief executive officers as well as other senior corporate officials; compared the economic performance of companies with and without profit-sharing plans; and examined trends regarding the introduction of plans.

I believe that the evidence is compelling. The data presented clearly show that organizations with profit-sharing plans turn in superior performances compared to those without.

A further point, however, remains to be made. The decision of whether to institute profit sharing should not be based on economic factors alone. In many ways, this decision is a philosophical choice by the owner/manager; it is a leap of faith.

---

[18] Richard J. Long, *Employee Profit Sharing and Share Ownership in Canada* (Saskatoon, Sask., 1991), 15.
[19] Long, 12–13.
[20] See Chapter 1, page 4.

4
CONDITIONS NEEDED FOR PROFIT SHARING

A number of conditions should exist before a company introduces a profit-sharing plan. If you are considering developing a plan, you or your advisers should objectively assess the situation to determine whether these conditions exist. These optimal conditions are as follows:

- good employee relations
- externally competitive base salaries
- internal equity
- profits
- reasonable organizational structure
- lack of other major changes that the plan is not able to support, i.e., stability
- good communications
- management commitment

Let's now look at each condition in greater detail.

EMPLOYEE RELATIONS

"Good" employee relations is perhaps better defined as "reasonable" employee relations. Although employee relations do not have to be spectacularly good, they should not be excessively bad either. Somewhere in the middle or better is an acceptable foundation for the introduction of a profit-sharing plan. Unfortunately, many employers mistakenly believe that a profit-sharing plan is the only solution to a poor relationship between owner/managers and employees. This is simply not the case. Profit sharing contributes to good relations or to improving mediocre ones but it cannot, by itself, solve all the problems involved in a very bad situation. Profit sharing supplements good management practices but is not a substitute for them.

Determining the state of your employee relations is often more difficult than it may appear at first, especially for an owner who is personally involved in managing the company. Such persons frequently view their company as "one big happy family," when this is not the case. They simply cannot be objective about their own management style. If they have a particularly strong personality—and founders of successful companies usually do—the employees are extremely unlikely to provide accurate feedback. In such a situation the use of outside advisers is usually advisable—not necessarily for any other aspect of the design of a profit-sharing plan but just for this reason.

These advisers could be professional management consultants or your accountants, lawyers, suppliers and customers. Customers, in particular, are in a good position to assess the situation because they deal with many of your employees, such as customer service, shipping, sales and accounts receivable. They also have little or no reason to "hedge their bets." Suppliers also deal with a large number of departments, such as purchasing, accounts receivable, receiving and warehousing, production scheduling and traffic management, although they may be somewhat more reticent than customers. Lawyers and accountants typically are skilled at delivering both "good" and "bad" news and should be able to provide objective feedback on the state of your human resource environment.

Another option is to use commercially available employee attitude or opinion surveys. Many organizations provide these surveys, especially companies that publish training and development materials. Although they vary widely, such surveys can measure issues such as employees' opinions about their current wages, salaries and benefit packages; their level of trust in management; the degree to which they feel they have the tools (money, authority, training) to do their jobs; and the general state of employee relations. As you will see below, all of these elements are important in assessing the feasibility of introducing a profit-sharing plan.

EXTERNALLY COMPETITIVE BASE SALARIES

The second essential condition is a program of base salaries or wages that is both externally competitive and internally equitable. If a company is externally competitive, the compensation package is comparable to that for similar jobs in other companies. For example, if your company pays electricians $15.00 per hour and a company in the same business just down the street pays them $20.00 per hour, you would not generally be considered externally competitive. However, if your electrician rate is $19.75 per hour, this would usually be viewed as competitive. You do not need to pay exactly the same amount as the competition to pass this test.

Relationship with the labour market can be measured in terms of just base salary, salary plus benefits, or salary, benefits and other compensation such as profit sharing. Although this issue can become complicated, for the purposes of deciding whether it is appropriate to proceed with a profit-sharing plan, you should probably concentrate on base salaries or wages.

Such information can be difficult to obtain but is often available from sources such as boards of trade and chambers of commerce (e.g., Board of Trade of Metropolitan Toronto), professional associations (Association of Professional Engineers of Ontario), industry associations (Canadian Manufacturing Association), government (Statistics Canada) and management consulting firms that specialize in the field of compensation. Alternatively, you can conduct your own survey or engage a consultant to do one for you. However, conducting your own survey is often time-consuming and difficult and can be expensive. For these reasons, the use of published surveys is vastly preferable. For the purpose of establishing competitiveness when installing a profit-sharing plan, these surveys are usually sufficient. A partial list of surveys available in Canada is provided at the end of this chapter and see Appendix 1 for additional sources.

INTERNAL EQUITY

Internal equity is a distinctly different but related concept. It is most important if you are considering basing the allocation of the profit-sharing fund on earnings (see Chapter 7 regarding the design of an allocation formula). Internal equity exists when two or more jobs that are considered of equal or comparable value are paid the same except for reasons of merit or seniority. The relative value of jobs is determined by the use of a process called "job evaluation," which is defined in the compensation literature as "the process of determining the relative value of jobs within the organization." It is extremely important to remember that job evaluation measures the value of the job itself and not the performance or the qualifications of the incumbent.

For example, suppose that an electrician is paid $15.00 per hour, while a maintenance mechanic is paid $18.00 per hour. The jobs have been "evaluated" and have been determined to be equally valuable to the company. Both incumbents have been on the job for more than five years and are considered to be satisfactory performers. In this situation, internal equity does not exist.

However, if the mechanic has been on the job for five years but the electrician has only one year of experience with the company, or the mechanic is considered to be a significantly better worker than the electrician, then internal equity exists as long as the electrician has the same opportunity to progress to the same rate as the mechanic.

PROFITS

Profits are the third important precondition necessary to introducing a profit-sharing plan. Although this may appear obvious, two important principles are involved.

The first principle involves financial planning. You should examine your strategic plans and/or any financial projections for the next few years and determine the level of profits that you can reasonably anticipate. That is, there should be a realistic possibility that when the plan is introduced, you will have some profits to share.

This does not mean, however, that profits must always exist to be divided among the employees. Many companies that consider installing profit-sharing plans seem to believe that employees will become "turned off" if there are times when the profits to be shared are low or non-existent. Yet this situation may well arise. However, with a good communications and educational program about how profits are achieved, employees should come to understand that fluctuations in profit levels are an economic fact of life. Some companies manipulate their calculation of profits so that employees almost always receive something. Unfortunately, this practice shelters employees from economic reality—losing the opportunity to make them true "partners" with management.

The other important aspect of profitability is the initial size of the potential profit pool (determining the exact definition of profits is part of Chapter 5). Your projections should show the potential to have a profit pool that constitutes from 3 to 5 per cent of the salaries or wages of the potential members of the plan. Psychological studies have shown that any payment has to be this size in order to be noticeable to the employee. That is, an amount smaller than 3 to 5 per cent of base pay will have little or no impact. Aim for this size of payment for at least the first one or two years of the plan.

ORGANIZATIONAL STRUCTURE AND JOB DESIGN

The design of jobs (the classic definition of a job is a collection of duties that can be performed by one individual) and their combination into various forms that can be put on an organization chart is another variable to be considered. As part of your preliminary analysis, you should examine these issues and ask: Is the organization of work at both the micro and macro level essentially reasonable and rational? This does not have to be a detailed and extensive organization study; it is essentially a broad review of the company structure. Indicators of problems in this area can include disputes about responsibility for specific activities; obviously overlapping responsibilities; grievances about job classifications; and essential functions that are not clearly defined or assigned to specific employees. If you have written job descriptions for all jobs, they can be reviewed as part of this analysis. Just the fact that job descriptions exist is a good sign. Lack of job descriptions is often an indicator of poor organization design.

STABILITY

Introducing a profit-sharing plan into an organization can be a major undertaking. It can utilize significant amounts of both management and non-management time and effort. It is therefore important that you do not start designing a profit-sharing plan when other major organizational change efforts are under way. Employees can only handle so much upheaval in their work lives.

Examples of such activities could be the building of, and movement to, a new office facility or the installation of a new manufacturing process requiring extensive retraining of the employees or even the hiring of many new staff members. Similarly, major downsizing is another condition that would make the introduction of a profit-sharing plan inappropriate at that time. A change in ownership, such as when family companies are sold by the second or third generation to large conglomerates or venture capitalists, is another such example. Other possible distractions include the introduction of such innovations as telecommuting, flexible working hours or a compressed work week.

The exception to this rule is where the profit-sharing plan is being advocated in order to support certain change efforts such as a Total Quality Management program or an ISO 9000 certification process.

Although this is obviously a difficult judgement call, you should carefully examine your objective(s) for the profit-sharing plan.

COMMUNICATIONS

Good communication is such a vital component of establishing a profit-sharing plan that Chapter 11 is devoted to this topic. A reasonably well-established program or system of regular communication of information from management to employees is critical. Furthermore, employees must believe in the integrity of management.

MANAGEMENT COMMITMENT

The final requirement before starting on the design of a profit-sharing plan is management commitment. This commitment can take two forms.

First, a profit-sharing plan requires additional management effort besides the effort involved in designing and implementing the plan. Some executives mistakenly believe that a plan can be introduced and then left to work its magic without any further effort on their part. Nothing is further from the truth. All types of plans—but especially non-cash plans—require ongoing administration to make them work. Necessary activities include the following:

- committee meetings
- preparing and delivering communications to employees
- accounting administration
- investment of deferred funds
- payouts to employees who leave the plan (as well as keeping track of them)
- dealing with suggestions by employees about ways to improve profitability

This final activity requires the most commitment from management because profit-sharing plans that significantly affect the performance of the company do so because they *involve* all employees. The power of employee involvement makes the difference. It is not just the fact that employees receive more money in the form of a share of the profits.

However, this creates a dilemma for many employers, especially in smaller companies where the chief executive officer is also the founder, owner and/or majority shareholder. Such entrepreneurs frequently find it difficult to accept employee involvement or participation in decision making. Since such entrepreneurs are, by definition, accustomed to acting independently, making decisions, and generally being the catalyst for all actions taken within or by "their" company, it can seem impossible to convince them of the value in allowing other people to be involved in managing the company.

In these cases, the following situation can develop. The company introduces the profit-sharing plan. The plan is based on the fact that management/owners and employees now have a common interest—they share in the success or failure of the enterprise. Since they share the results, employees assume that they should share in the running of the organization. Yet this is where the plan can fall apart. Many managers/owners have not thought beyond the sharing of results to the sharing of power—which is really what the plan is all about. Otherwise, the plan is just another mechanism to give away money.

Employees are initially motivated to contribute to the company's success and they begin to offer suggestions for improvement; to critically observe traditional, standard operating procedures; and even to challenge management decisions. Even though their comments usually have at least some validity, since they are critical of management (usually constructively but not always), managers often find it difficult to accept the comments offered. Anger and rejection of the comments or suggestions are frequently the responses. This is a critical point.

In this way, employees receive two messages. First, they believe that they are participants in the success of the organization, and, second, they believe that their input is unwelcome. They may try one or two more times to contribute, but, if they receive the same response, their involvement will cease. If this happens, the most powerful component of a successful broad-based employees profit-sharing plan is lost. The plan will become just another part of the basic compensation package and a different administrative system used to distribute money.

The characteristic of profit-sharing plans that has the most effect on the performance of the organization is "the power of the common person." A properly constructed and administered plan will motivate the average employee to contribute in ways that will frequently astound the owner/manager. Since employees are involved, they will tend to "work smarter" rather than just harder.

One of the main things that employees want from their job is a sense of involvement. This can be accomplished simply by asking employees for their opinion and then paying attention to their responses. You cannot fake employee involvement and you must be prepared to make drastic changes in your style of management to make the profit-sharing plan successful. Otherwise, the plan just becomes a mechanistic formula for paying out more money.

LACK OF PROPER CONDITIONS

What happens if these conditions do not exist? This supposedly simple question can cover a wide range of possibilities; the company could be slightly deficient in one condition or it could be seriously lacking in most of them. Although this is a judgement call, you should not develop a profit-sharing plan without being convinced that conditions are favourable. If you are not convinced, work on creating the proper environment and defer the development of the profit-sharing plan.

CONCLUSION

You should carefully examine your company, and yourself, before taking even the first step towards involving employees in the design of a profit-sharing plan. You are ready to proceed if the following conditions are in place:

- your profits for the next two or three years meet the 3 to 5 per cent of payroll test;
- relations with your employees are reasonable;
- your base salaries and wages are externally competitive and internally equitable;
- your communications systems are working effectively; and
- you, the owner or chief executive, are prepared to change your style (if necessary) so that employees can get involved and participate in the running of the company.

LIST OF PUBLICLY AVAILABLE SALARY, WAGE AND BENEFIT SURVEYS IN CANADA[1]

The following organizations provide either periodic surveys of wages, salaries or benefits and/or customized analyses of labour market data. Refer to Appendix 1 for other possible sources such as human resources associations.

THE ASSOCIATION OF PROFESSIONAL ENGINEERS OF ONTARIO

Telephone: (416) 224-1100
Fax: (416) 224-8168
A.P.E.O. produces two annual surveys for Ontario:
- Ontario Engineers Salaries–Survey of Members
- Ontario Engineers Salaries–Survey of Employers

BOARD OF TRADE OF METROPOLITAN TORONTO

Telephone: (416) 366-6811
Fax: (416) 366-4906
The Board offers the following annual surveys that cover all industries in the Greater Toronto Area:
- Data Processing
- Executive
- Middle Management
- Professional and Supervisory

BUREAU OF LABOUR INFORMATION,
HUMAN RESOURCES DEVELOPMENT CANADA

Telephone: 1-800-567-6866
Fax: (819) 953-9582
Wage Settlements Bulletin published monthly; reports on negotiated settlements in both public and private sectors.

CANADIAN ADVANCED TECHNOLOGY ASSOCIATION

Telephone: (613) 236-6550
Fax: (613) 236-8189
The Association conducts an annual, national survey of compensation and benefits for Canada's advanced technology industry. The survey is available only to participants.

---

[1] This list includes only surveys produced by not-for-profit organizations. However, many accounting and/or management consulting firms produce their own surveys that cover an extremely wide variety of industries, occupational groups, geographical areas and compensation components.

CHAPTER 4 — CONDITIONS NEEDED FOR PROFIT-SHARING

CANADIAN MANUFACTURERS' ASSOCIATION

Telephone: (416) 798-8000 Extension 223 OR 1-800-268-9684
Fax: (416) 798-8050
The Association conducts the following surveys, by province, annually: (manufacturers only). Customized surveys are also available.
- Compensation and Benefits Survey for Salaried Employees
- Compensation and Benefits Survey for Hourly-Paid Employees

CITY OF MISSISSAUGA (ECONOMIC DEVELOPMENT DEPARTMENT)

Telephone: (905) 896-5016
Labour and employment statistics for the City of Mississauga.

GRAND VALLEY HUMAN RESOURCES PROFESSIONALS ASSOCIATION

Telephone: (519) 744-6541 (The Letter Service)
Fax: (519) 744-2038
The Grand Valley and Guelph and District Human Professionals Associations jointly produce the Annual Compensation and Benefits Survey, which covers most occupational groups except for executives. Includes association members in the Kitchener-Waterloo, Guelph, Cambridge, Brantford, Fergus and Stratford area. Available to non-members.

HUMAN RESOURCES DEVELOPMENT CANADA–HALIFAX

Telephone: (902) 426-2902
Fax: (902) 426-7641
Produces Wage Rates and Conditions of Employment Report twice a year with regular updates which shows wage ranges for approximately 250 occupations as well as working conditions. It covers both the public and private sector in Metropolitan Halifax.

INSTITUT DE RECHERCHE ET D'INFORMATION SUR LA REMUNERATION

Telephone: (514) 288-1394
Fax: (514) 288-3536
The Institute conducts an annual survey on salaries, benefits and working conditions for major sectors of the Quebec economy. It also does a wide variety of special studies related to compensation.

LONDON CHAMBER OF COMMERCE

Telephone: (519) 679-9160
Publishes an annual Compensation Survey, which includes 35 salaried, hourly, supervisory and sales jobs. It also includes benefits.

32 PROFIT SHARING IN CANADA

ONTARIO MINISTRY OF LABOUR,
OFFICE OF COLLECTIVE BARGAINING INFORMATION
Telephone: (416) 326-1260
Fax: (416) 326-1277
The Office maintains a library of over 10,000 collective agreements and can do customized, computer searches of various provision of agreements.

STATISTICS CANADA
Telephone: (613) 951-4090
Fax: (613) 951-4087
Conducts a wide variety of analysis of labour earnings and hours of work; contact your Regional Office.

UNITED STATES BUREAU OF LABOR STATISTICS
Telephone: (202) 606-6288

VANCOUVER BOARD OF TRADE, WESTERN COMPENSATION & BENEFITS CONSULTANTS
Telephone: (604) 683-9155
Fax: (604) 687-2315
The following annual surveys are available for all industries in the Greater Vancouver Area:
- Clerical Salary Survey
- Information Processing Salary Survey
- Report on Employment Practices, Employee Benefits, Salaries
- Administration Practices, Perquisites and Other Cash Compensation
- Middle Management Salary Survey
- Marketing and Sales Salary Survey

WINNIPEG CHAMBER OF COMMERCE
Telephone: (204) 944-8484
Fax: (204) 944-8492
- Clerical Salary Survey–Public Sector
- Clerical Salary Survey Results–Private Sector
- Clerical Salary Survey Results–Non-Profit Associations

Covers Winnipeg area; usually done annually; last done in 1993.

5
GENERAL PRINCIPLES OF DESIGNING A PLAN

INTRODUCTION

This chapter discusses issues that are common to all types of plans (cash, deferred, etc.), must be decided at a very early stage of the design process, and are exclusively the prerogative of management. These issues are as follows:

- the objective(s) of the proposed plan;
- the type of plan (subject to a qualification);
- the nature and degree of employee involvement in the design process;
- announcing the project to the employees;
- the use of "copy-cat" plans;
- choosing a start date for the plan;
- planning the first meeting of the profit-sharing committee; and
- the definition, and amount, of profits that the company will contribute to the plan, including the issue of the level of pay

that you wish to achieve relative to your previously identified labour market.

CHOOSING AN OBJECTIVE

An extremely important step in the design of an employee profit-sharing plan is selecting an objective(s). This objective is the answer to the supposedly simple question, "Why do I (the employer) want to introduce a profit-sharing plan?" Unfortunately, plans are often implemented for reasons such as "it seemed like a good idea at the time," "my neighbour put one in her company" or "profit-sharing plans are trendy these days; everyone is doing one."

Lack of a clear objective for installing a profit-sharing plan can be a major obstacle to its successful design and implementation. The following are some common reasons for introducing a broad-based profit-sharing plan:

- provide an incentive for employees to increase productivity;
- provide retirement income (either in addition to, or in lieu of, a registered pension plan);
- reduce absenteeism;
- reduce turnover;
- improve ability to attract superior employees;
- reduce fixed costs;
- provide opportunities for increased income since one of the traditional methods, being promotion, has decreased in availability. This is because organizations have been "flattened," that is, the number of levels has been reduced and there are simply fewer positions to be promoted into. At the same time, the baby boomers are reaching the age where they would be eligible for promotions but, since there are so many of them, opportunities for all are sharply reduced.
- reinforce teamwork, whereas traditional pay systems have emphasized individual performance;
- provide additional benefits such as disability insurance;
- increase employees' current income;
- encourage employee saving;

CHAPTER 5 — GENERAL PRINCIPLES OF DESIGNING A PLAN

- provide an employee with a "stake" (ownership) in the fortunes of the company;
- improve competitive position, profitability, customer service, quality levels, etc.;
- reinforce the introduction of such business improvement innovations as Total Quality Management, re-engineering or ISO 9000; and
- provide employees with a "nest egg" for use in emergencies.

Your choice of an objective will influence virtually all decisions in the following process. Although this will become clearer as you progress through this book, two simple examples illustrate the point.

If the objective of the plan is to provide retirement income (in addition to, or instead of, a pension plan), the choice of type of plan will clearly be a DPSP rather than a cash plan. An objective of reducing turnover will require a shorter initial waiting period to join the plan than would be the case if turnover is not an issue since most turnover occurs in the first year or so of employment (see Chapter 6, "Membership in the Plan").

TYPE OF PLAN

A major issue that should be decided at the very early stages is whether to specify the type of plan (cash, deferred, etc.) or to allow employee input into that decision (employee involvement is discussed in the next section). The reason for this is simple: if employees are given a choice on this subject, they will almost always select a cash plan—especially younger employees. If, as the employer, you have no preference as to which type of plan is chosen, the process will have more credibility with employees if they are given the opportunity to express their opinion on this issue.

If, however, you have a preference, make your decision and advise the employees accordingly. This is especially recommended if you wish to have any plan other than a cash plan. There is no point in pretending to allow the employees input into a decision when you know how they will respond and you have already decided.

EMPLOYEE INVOLVEMENT

After choosing the objective(s) and type of plan, the next major task is to determine if, and how, employees will be involved in the process.

Basically, you have two choices. You can design the plan alone—possibly with a few close advisers or consultants—or you can involve employees at all levels in the process. As usual, both approaches have advantages and disadvantages, although I clearly favour the second alternative.

The advantages of the first approach are primarily speed and convenience. The main disadvantage is that employees will be significantly less understanding and supportive of the plan when it is installed. Most of those employers reading this book who head privately owned companies and consider their employees almost as "part of the family" will inevitably believe that this does not apply to them. Unfortunately, however, it does. I have seen it happen with clients who, by any modern standards, would be considered enlightened employers. They were also not unionized. The issue being raised is not union versus non-union. It is basic human nature to suspect and/or resist change.

Regardless of how good an employer you are, you will have a better profit-sharing plan if you involve employees in the design process. The technical improvements that will result, although reasonable, will be insignificant compared to the improved understanding and acceptance of the final product.

How do you involve employees? The best way is to establish a profit-sharing committee that can either be elected or appointed. It is probably better to appoint the first committee; the reasons will be obvious from the discussion below about the characteristics of such a committee.

The ideal size of a profit-sharing committee is from five to 10 persons, although this number may vary in order to achieve the other desirable characteristics. Ideally, members of a committee should be:

- representative of all sections, levels, departments and divisions of the organization;
- a cross-section of the demographic profile of the company in terms of age, seniority, gender, education, ethnicity, etc.;
- considered "opinion leaders" by their respective constituency;

they should be sufficiently well respected that their peers will feel comfortable being represented by them;
- willing to act as spokesperson for their group;
- reasonably articulate—they must be able to express the opinions and concerns of the group of employees they are representing (this can often be a problem if a large proportion of the employees are not fluent in one of our official languages); and
- be capable of understanding the concepts involved.

Obviously, not all these conditions can be met. Furthermore, do not be afraid to select an employee who may have been a "thorn in your side" if he or she meets the other criteria. The act of involvement often turns these employees into converts and greatly increases the credibility of the process with the other employees.

You should also determine whether you want to select the committee yourself or have committee members elected by their respective departments. In my experience, I have found it better to select the initial committee. There are several reasons for this, the most important of which is that it enables you to see that the committee meets the above criteria as closely as possible. Open elections are unlikely to achieve this objective to the same extent. Elections can always be held later as part of the ongoing administration of the plan. In fact, one of the issues that the first committee will be asked to address is the selection process for members of future committees (see Chapter 12, "Administration").

Prospective committee members should be asked if they wish to volunteer for the assignment. If an employee meets the profile but does not wish to participate, another candidate should be found.

A variation on the option of management selecting the committee is to generally advertise the need for committee members and to ask employees to volunteer. Although this "advertising" could be done informally by word-of-mouth, a more effective way is to formally post the job (of committee member) on bulletin boards or to mail it to employees' homes. This posting should include a brief description of the responsibilities of the committee members and the qualifications listed above. Management can review the applicants to ensure that they meet the required qualifications and make the appropriate selections.

Potential committee members should be assured that all time spent in committee work will be treated as though it is regular work time and there will be no loss of pay, overtime, seniority, etc. Committee responsibilities should be clearly determined by management before the first meeting. The question of a cash or other type of plan has already been addressed and should be communicated to the committee. The other major issue is whether the committee can decide on the final form of the profit-sharing plan or just make recommendations to management or the owner of the company. Generally speaking, the CEO or the board of directors will reserve the right of final approval.

If there is more than one major location with significant numbers of employees, it is possible to have one committee for each location with management or external advisers creating the common ground or to have an overall committee representing all locations. Costs for travel and employee time are the major constraints if you decide on a two-tier committee system.

ANNOUNCEMENT

Early in the proceedings there should be an announcement to all employees from the chief executive officer regarding the decision to develop a profit-sharing plan and the process that will be followed. This announcement should include:

- the objective(s) of the plan;
- the type of plan chosen or the fact that this decision will be part of the process;
- the names, job titles, and locations (departments) of the committee members;
- how the committee was chosen (i.e., representatives of all major groups, departments, levels, etc.);
- the name of the chair of the committee;
- whether the committee will make the final decisions or will make recommendations to management; and
- the name of a contact person (other than committee members) if any employee has questions; this is usually the manager of human resources.

An example of an announcement letter is shown in Exhibit 1.

CHAPTER 5 — GENERAL PRINCIPLES OF DESIGNING A PLAN

EXHIBIT 1

SAMPLE ANNOUNCEMENT MEMO FOR A PROFIT-SHARING PLAN

TO: All Employees of XYZ Company
FROM: John Smith, Chief Executive Officer
SUBJECT: Employees Profit-Sharing Plan

I am pleased to announce that I (we, the executive committee, senior management, the owner) have decided to develop and implement an employee profit-sharing plan. The objective of the plan is to _____.

The plan will be a tax-sheltered deferred profit-sharing plan (DPSP).

To assist us in developing this plan, I have established a profit-sharing committee. The members of the committee were selected so that they would comprise a representative cross-section of the company. The following employees have agreed to serve on this committee:

George Scrimger:	Human Resources (Chair)
Sally Brown:	Quality Control
Ian Babcock:	Production
Peter Harris:	Finance
Angela Strathdee:	Marketing
Allen Morgan:	Field Sales
Jim Johnston:	Warehouse

The purpose of this committee will be to learn about the issues involved in the design of a profit-sharing plan and to obtain your opinions on each issue. Committee members will provide a summary of the opinions of their department or group to the committee, which will then try to develop a consensus that can be presented to senior management. Final details can then be worked out.

I encourage you to discuss the issues thoroughly with your committee representative. The next meeting of the committee is scheduled for _____ so please try and meet with them either individually or in groups before then.

If you have any questions about this project, please contact me or George Scrimger, Manager, Human Resources. I would like the plan to be in effect for the fiscal period (year, month, quarter) starting _____.

(signature)

John Smith
Chief Executive Officer

COPY-CAT PLANS

Using "copy-cat" plans is the process whereby the owner/manager contacts several companies that have profit-sharing plans—supposedly successful ones—and obtains copies of these plans. The owner/manager then reviews them and selects and installs the most appealing one, with little or no change.

Many companies take this approach since it has the enormous advantage of being very quick and seems to save the company a lot of time and effort by not "reinventing the wheel." However, this approach does pose two significant disadvantages.

First, it imposes a profit-sharing plan on the organization that was developed for a different company, often in a separate industry, with a different culture, group of employees, profitability, objective(s) and at another time. I believe that it is asking too much to expect a copy-cat plan to meet the needs of your organization without an extensive review and overhaul. Such a review, of course, negates the time and effort savings mentioned above.

Second, you cannot use the employee involvement process that I have advocated above and receive the benefits thereof by using a copy-cat plan. These benefits are formidable and it can be a major mistake to forego them.

FIRST MEETING OF THE PROFIT-SHARING COMMITTEE

The general procedure to be followed by the committee is based on the assumption that the members will act as representatives of the areas they work in and as conveyors of information from these areas to the whole committee and vice versa.

A typical agenda for the first meeting of the profit-sharing committee should include the following:

- introductions—all members may not know each other;
- remarks by the chief executive officer explaining the decision to develop a profit-sharing plan;
- general introduction to profit-sharing plans;
- the major issues and options to be addressed by the committee;
- introduction and role of the chair;
- rules of conduct for the committee;

CHAPTER 5 — GENERAL PRINCIPLES OF DESIGNING A PLAN

- date for the next meeting; and
- any other business.

During this first meeting, committee members will learn about the major issues in the design of a profit-sharing plan. A member of management or an outside adviser such as a consultant will be responsible for providing this information. This individual should probably also chair the committee. Who should this person be? I believe that it should not be the CEO or the owner/manager since his or her presence will often intimidate the committee members, especially those new to the company and/or from lower levels of the organization. If the company is large enough to have a human resources manager, then this individual is often the ideal chair. Otherwise, the most senior-level person on the committee can fulfil this role, provided that he or she has the facilitation skills necessary.

Committee members usually have many questions at this time and they typically expect answers. However, do not provide answers. You are not yet in a position to do so since you have already committed to a process of employee involvement; the only appropriate action at this point is to provide the questions and alternatives.

The chair of the committee should allow committee members to ask as many questions as they wish. If you limit the discussion, it may appear that management has something to hide. Remember, committee members—and employees in general—will continually ask questions such as "Will the plan include _____?" If you are truly committed to the process of employee involvement, your only legitimate answer is that you do not know. Such questions will be tentatively answered when the committee meets again and reports on other employees' opinions on various issues.

Another aspect of the role of the committee members is how they conduct themselves during committee meetings. In other words, do they follow the rules of conduct for the committee? I suggest the following guidelines will turn the committee into a very effective team:

- all members should attend all meetings and arrangements should be made to ensure that interruptions, such as telephone calls, should be minimized;

- everything is open for discussion;
- all discussion is confidential: nothing leaves the committee room;
- all disagreement is constructive and not confrontational; and
- everyone should contribute to the discussion.

Once you have discussed all the issues surrounding the role of the committee members, you can proceed to the design issues that apply to all types of plans: membership, allocation, communication and ongoing committee membership and responsibilities. These issues are discussed in the following chapters, as are other issues such as vesting, which apply only to non-cash plans.

After this initial meeting, committee members return to their respective departments and pass on this information to their colleagues. You may wish to provide written material summarizing the issues and options. Checklists are provided in Chapter 13, which can be photocopied and provided to the committee. They should be directed to discuss the issues with as many of their fellow employees as is practical. Some committee members will choose to have a meeting of the whole group that they represent; others will have one-on-one discussions; some will even ask for written responses after distributing prepared material. Allow committee members to choose how to circulate the information. The only major logistical issue that management must address is the disruption of the work flow. I suggest that you be as accommodating as possible even to the point of shutting down a production line or paying employees overtime to stay after regular working hours.

Setting the date for the next meeting involves a little more effort than usual. You should ensure that each committee member has sufficient time to contact all of his or her constituency. Consider events such as vacation shutdowns, inventories, year-ends, and the Christmas season, as well as the number of employees that the committee member is representing. In most cases, two or three weeks is usually sufficient time. It is equally important that the time period not be so long that the momentum of the process is lost.

CHAPTER 5 — GENERAL PRINCIPLES OF DESIGNING A PLAN   43

EMPLOYER CONTRIBUTIONS TO THE PROFIT-SHARING PLAN

This section addresses the major question of the size and nature of the contribution that the company will place in the profit-sharing pool for distribution to the employees. Although you may decide otherwise, this issue need not be discussed with the committee; it is strictly an owner/manager decision. The five issues involved here are

- definition of profits;
- portion of profits;
- type of formula to be used;
- timing and frequency of the calculations and payments; and
- overall compensation policy of the company.

DEFINITION OF PROFITS

Profits can be defined in two ways: net profits before and after taxes. The 1990 Hewitt Survey found that 82 per cent of companies surveyed use before-tax profit, while the remainder use after-tax profit.[1] A related question is whether to include extraordinary gains and losses (such as the sale of a building) and investment gains and losses. If employees have a major influence on these elements, the results should be included and vice versa. This is obviously a judgement call. Further considerations include the effect of any payments on the corporation's return on investment/assets and cash flow.

TYPE OF FORMULA

Another issue is whether to use a specified or "declared" formula (e.g., 5 per cent of net profits before tax) or to vary the formula from year to year at the discretion of the company (i.e., to use a "discretionary" formula).

Declared formulas have two variations: a straight percentage of all profits (15 per cent of net profits after taxes) or a percentage of only those profits above a specified minimum (e.g., 50 per cent of all profits after the first $100,000). The Hewitt Survey showed the following distribution in 1989[2]:

---

[1] Hewitt Associates, *Survey of Canadian Profit-Sharing Plans* (North York, Ontario, 1990), 10.
[2] Hewitt Survey, 9.

FORMULA	PERCENTAGE OF COMPANIES
Straight percentage of all profits	27
Percentage in excess of a minimum	21
Percentage of participants' pay	20
Employee contributions	8
Discretionary	15
Service	2
Other	7

A declared formula can also be graduated so that the percentage increases as the level of profits increases, for example:

PROFIT LEVEL	PERCENTAGE OF PROFIT
first $100,000	5
next $400,000	7½
next $500,000	10
all profits over $1,000,000	15

The disadvantage of the graduated approach is that it is somewhat more difficult to communicate the results of the plan to employees. The advantage, however, is that the company is assured a minimum return on investments or assets.

Although discretionary formulas are used by 15 per cent of companies, they are probably contrary to the spirit of profit sharing in that there is minimal commitment to the employees. In most discretionary plans, the company only decides the percentage of profits to allocate after the end of the fiscal year. This type of formula is also in conflict with the major theory of using pay to motivate employees because employees do not know what their reward will be if they do produce a high level of profits (see Chapter 16). In fact, they are not certain that there will be any payout at all.

AMOUNT

Once you have defined which profits (e.g., net before taxes) you will share with employees, you must decide what portion of these profits will be contributed to the profit-sharing pool. Aside from

CHAPTER 5 — GENERAL PRINCIPLES OF DESIGNING A PLAN   45

the usual financial considerations, there are two main approaches to making this determination: as a percentage of total company profits and as a percentage of payroll. The first approach will be the final formula (called the "contribution" formula) to be used on an ongoing basis; for example, 12 per cent of net profit before taxes. The second approach is a test of the first.

I examined the sample plans provided in *The Book of Plans*, which was published by the Profit Sharing Council of Canada.[3] The distribution of employer contribution percentages is shown in the following table:

ANALYSIS OF PROFIT SHARING COUNCIL'S *THE BOOK OF PLANS* REGARDING PERCENTAGE OF PROFIT ALLOCATED TO FUND

PERCENTAGE OF ALL NET PROFIT BEFORE TAXES	NUMBER OF PLANS
3–5	1
4–6	1
5	1
6	2
6.75	1
7	4
8	4
8–10	1
10	23
11	1
12	3
12.5	2
13.5	1
14	1
15	11
20	9
22.5	1
25	5
27	1
All Others*	57
Total	130

*This category includes all those plans that do *not* use a percentage of net profit before taxes. It includes discretionary formulas, a percentage of profits after a predetermined minimum profit or return on investment, and those that did not report a formula.

[3] *The Book of Plans* (Don Mills: The Profit Sharing Council of Canada). The council was an association of companies with broad-based employee profit-sharing plans. It ceased operations several years ago.

The second method of determining how much profit to contribute to the plan is to calculate any proposed amount as a percentage of the total cash payroll. Cash payroll excludes benefits and perquisites. The reason for taking this approach is that employees tend to assess the appeal of any possible payment as a percentage of their own pay. As discussed earlier, a target guideline of 3 to 5 per cent of base pay is considered appropriate. However, many plans in Canada operate successfully and seldom pay more than 2 per cent of pay. Any amount less than this becomes too small to affect employee behaviour. That is, it becomes unnoticeable. According to the Hewitt Survey, the average payment in Canada in 1989 was 7.96 per cent of pay. In 1988, the average payment was 9.44 per cent and was 7.96 per cent in 1987. The average payment in 1989 as a straight percentage of profits was 5.7 per cent.

In determining the amount of profit, the company must consider both approaches even though they may not be compatible. If you are unsure which level of contributions you can sustain, it is better to start with a lower percentage and then increase it later when the fiscal performance of the company becomes more clear.

Frequency and Timing of Payments

Another issue is selecting the frequency of payments. In Canada the vast majority of employers make an annual contribution based on the results of the whole fiscal year. However, some companies make quarterly, semi-annual or even monthly payments.

Two of the salary surveys by the Board of Trade of Metropolitan Toronto show the following distribution of payments from profit-sharing plans for the respective employee groups.

SURVEY	PAYMENT FREQUENCY		PERCENTAGE OF COMPANIES	
	ANNUAL	SEMI-ANNUAL	QUARTERLY	MONTHLY
Clerical*	82.3	7.6	5.1	2.5
Information Technology**	85.7	2.9	2.9	2.9

* Metropolitan Toronto Board of Trade, *Clerical Salary Survey*, 1994, 47.
** Board of Trade, *Information Technology Salary Survey*, 1994, 53.

CHAPTER 5 — GENERAL PRINCIPLES OF DESIGNING A PLAN

Annual calculations have the advantage that they are probably more indicative of the company's performance since many actions do not have any immediate effect. However, more frequent calculations mean that the payment is closer to when the action that resulted in the profit occurred, and therefore it has more of a direct motivational effect.

In terms of the timing of payments, they are usually made after the receipt of audited financial statements—typically three to four months after the fiscal year-end. However, some companies make an interim payment (e.g., an estimated 50 per cent of the total payment) with the rest to follow at a later date. S.C. Johnson Wax of Brantford, Ontario, has used this approach for many years. Their fiscal year ends on June 30, at which time they pay out an estimated 45 per cent; the remainder is paid at Christmas. This approach gives staff an immediate reward at the exact end of the fiscal year without any danger of making the wrong payment. It also provides employees with extra cash for summer vacations and at Christmas.

COMPENSATION POLICY

The final issue affecting the size of the employer contribution to the profit-sharing plan is the company's compensation policy. This means that you have to determine how well you want to pay your employees relative to your perceived labour market. To do this, you must make two policy decisions.

First, you must decide how competitive you want to be in terms of base salaries or wages. However, with a profit-sharing plan in addition to base pay, you have some latitude regarding your degree of competitiveness. For example, if the average pay in your industry is $15.00 per hour, you could decide to pay at this level or, at some level below that, and to use the profit-sharing plan to make up the difference. Several examples can illustrate this approach.

	INDUSTRY AVERAGE	PROPOSED BASE RATE	+	PROFIT	=	TOTAL
A	15.00	14.00	+	1.00	=	15.00
B	15.00	14.00	+	1.50	=	15.50
C	15.00	13.00	+	3.00	=	16.00
D	15.00	15.00	+	1.00	=	16.00

In Option A, your total cash pay is equal to the industry average. However, this approach may make it difficult to attract and retain workers since part of the competitive pay package is "at risk." In other words, there will inevitably be years in which there is no profit sharing and your cash package will not be competitive. The standard response to this dilemma is to make the profit-sharing payments, under "normal" circumstances, larger so that the total package is higher than the industry average. Option C is a more extreme illustration of this than Option B. The advantage of all three approaches is that your fixed costs are always less than those of your competitors. Option D illustrates the more conventional approach, which is to add profit sharing to competitive base salaries. However, the other approaches are becoming more common in the 1990s. You should carefully examine your own values, and your objectives in establishing a plan, before deciding on your policy position.

6
MEMBERSHIP IN THE PLAN

This chapter discusses the next major issue that is part of all types of profit-sharing plans, both cash and non-cash—membership in the plan.

Membership simply means "who is in the profit-sharing plan?" Since the definition of broad-based profit-sharing plans includes virtually all employees, it is only necessary to specify the limits of membership.

Membership can be defined in five ways:

- category of employment, i.e., full-time, part-time, temporary, occasional, etc.;
- length of service;
- when, and under what circumstances, employees leave the company;
- union status; and
- whether a particular employee group is already receiving some form of contingent or variable pay such as sales staff on commission or production staff involved in a factory-wide incentive plan.

CATEGORY OF EMPLOYMENT

Most plans include all full-time employees, subject to a length-of-service requirement (e.g., to be eligible for a pay-out under the profit-sharing plan, an employee must have served one year). However, many companies employ part-time people. Should these employees be members of the plan? If they are "regular" part-timers, that is, if they work the same hours or days every week, they are often included in the plan. Even though they are not full-time, such employees are typically considered to be part of the organization and to make a contribution to the success (profitability) of the company. This is especially true in the retail industry.

These employees are different from "occasional" or "casual" part-timers (also called temporary), who are usually hired for specific periods or for the duration of a specific task such as the annual inventory. Students hired for the vacation period are also considered to fall into this ("occasional or "casual") category.

Those employees included in the plan who are other than full-time are usually given a share of the profits in the same proportion that their work week or year is to a full-time employee. For example, if an employee works two days a week in a company where the standard working hours are seven per day and 35 per week, the employee's share of the fund will be

$$\frac{\text{Employee hours per week}}{\text{Standard weekly hours}} = \frac{7 \times 2}{35} = \frac{14}{35} = 40\%$$

This employee will receive 40 per cent of what a full-time employee will receive.

If you decide to include regular, part-time employees, two other issues arise. The first is the effect on the allocation formula if length of service is part of it. This is discussed in more detail in Chapter 7.

The second issue is the length of service requirement. If the plan stipulates that an employee complete three months of service before becoming a member, you must define when the part-timer has completed three months. The usual approach is to admit the employee to the plan when he or she has accumulated the equivalent, in total, of the requirement for full-timers. In the example used above, that employee would require (assuming the service requirement is three months and using 4.3 as the average number of weeks in a month):

$$\frac{\text{Employee hours/week}}{\text{Standard weekly hours}} \times 4.3^* \times 3 =$$

$$\frac{2 \times 7}{35} \times 4.3 \times 3 = 5.16 \text{ months}$$

* There are an average of 4.3 weeks in a month.

Therefore, this employee will have to work 5.16 months before becoming a member of the plan.

LENGTH OF SERVICE

Length of employment is the other major criterion used to determine membership in profit-sharing plans. The two major sub-issues to consider are the length of time that it takes an employee to become an effective contributor to profitability and the period during which turnover is highest.

If, for example, there is very high turnover during the first six months of employment, it is effective to require that employees have at least six months' service before becoming members of the profit-sharing plan. This achieves two objectives: it reduces the amount of administrative effort required and ensures that only employees with at least a minimal commitment to the company are included in the plan. The most common waiting period is one year of service although one of the most famous plans in Canada—that of Dofasco—requires two years of service. Some combination plans have different eligibility periods for the cash and deferred portions of the plan (e.g., immediate eligibility for the cash plan and two years for the deferred portion). This recognizes different objectives for the different parts of the plan; immediate motivation (cash) versus long-term perspective (deferred). It also increases the administrative and communication aspects of the profit-sharing plan.

If you decide to use length of service as part of your membership criteria, you must decide exactly when to make the calculation. For example, if you choose a six-month eligibility period, employees can become members of the plan at two times during the year: at the end of the six months of employment and on the first day of the fiscal year following the year in which the employee completes six months of employment.

This is illustrated below:

Start date - April 1, 1995
Fiscal year - January 1–December 31

```
                         1995
Jan. 1      Apr. 1              Oct. 1      Dec. 31
```

If the first option is chosen, the employee will become a member of the plan on October 1, 1995. Under the second option, he or she will join the plan on January 1, 1996. The practical impact of this second option is to make the eligibility period longer. However, it does simplify administration because all new members join on the same date. This is not a major issue unless the company has an extremely high turnover rate.

If you choose the first option—which most companies do—you must decide on two other items. Will the membership be retroactive to the start of employment (April 1 in the example) or will it include just the period from October 1 to the end of the year? In either case, the employee will only be eligible for a pro-rata share.

EMPLOYEES WHO LEAVE THE COMPANY

Other considerations include employees who were normally full-time and were on the payroll at the start of the year (assuming that the plan makes annual calculations) but left the company before the end of the year because of death, retirement, long-term disability (illness), quitting or termination by the employer. The usual practice is to include those who left the company—other than those who quit or were terminated—in proportion to the part of the year that they were employed. For example, an employee who retired after 10 months of the fiscal year would be given ten-twelfths of what he or she would have earned if employed for the full year. In the case of illness, it is useful to separate short-term illnesses from long-term disability (assuming that the employer has LTD insurance). Unless there is an attendance component (i.e., actual hours worked forms part of the allocation formula), short-term illnesses typically do not remove the employee from membership in the profit-sharing plan.

Dealing with employees who chose to leave the company is a much more contentious issue. The argument for including them in the plan is that they have contributed while they were there (otherwise they would have been terminated) and are therefore entitled to a pro-rata share. On the other hand, it is often argued that by the employee leaving, the company has incurred various costs (replacement costs, lost productivity, etc.) and thus the employee should not receive a share. In discussing this issue, many people assume that employees who quit do so in anger or with a grudge against the company. This situation, however, is the exception rather than the rule; many quits are for legitimate reasons, often beyond the control of the employee, such as a return to school, the need to accompany a relocated spouse, or a parent's desire to stay at home with the children or to take care of his or her parents.

There is no "scientific" way to decide this issue; it is essentially a choice that reflects the basic beliefs and values of the committee members and the organization. One common compromise is to stipulate a minimum amount of time as a member of the plan for the fiscal year in question, such as nine months.

Employees who are laid off during the fiscal period must also be considered regarding the membership question. Although they are not working, they are still considered to be employed and are often still enrolled in benefit plans (this enrolment is usually the distinction between layoff and termination). The question is whether they should receive a pro-rata share for the period they worked, possibly subject to a minimum time period, or not be included at all in the distribution for that year.

The arguments for and against including employees who are terminated by the company for cause or lack of performance are essentially the same as for those who quit except that the issue of whether they contributed is much clearer. The two most common approaches are to exclude them from the plan automatically or to make the decision on a case-by-case basis. Often management will decide to terminate an employee in the eleventh month of the fiscal year, supposedly on the basis of performance, but without any prior warning. In such cases, it may be more equitable (and safer legally) to pay the employee a pro-rata share of the profits.

EMPLOYEES ON VARIABLE PAY PLANS

Whether to include employees who receive commission (these are usually sales staff) or other bonus programs is another difficult and often contentious issue. Such plans can create barriers and conflicts of interest for the group. There are two points of view on this issue. One group argues that since profit sharing is a form of incentive pay, commissioned sales staff already have an incentive plan and including them in the profit-sharing plan is duplication. They also argue that this is especially true if the allocation formula is partly or totally determined by employee earnings since increased earnings under a commission or bonus plan will result in further increased payouts under the profit-sharing plan.

The counter-argument is that a profit-sharing plan is supposed to unite and motivate all employees towards a common goal and excluding any group violates this concept. This group also argues that commission payments are really part of a salesperson's "regular" compensation package and that these employees should not be excluded from the profit-sharing plan. There is also the argument that excluding them means that there are two classes of employees: those in the plan and those who are not. Those who are not in the plan are therefore motivated towards different objectives than those who are included in the plan. The sales force is motivated by either increased or total sales, depending on the exact nature of the commission formula; there is little direct connection with profitability. In fact, one can argue that by excluding them, they could possibly be motivated to increase sales and decrease profitability.

In general, I believe that this second argument has more merit, and consequently I usually encourage employers to include commissioned sales staff in the profit-sharing plan. This choice will also have an impact (either positive or negative) on the allocation formula if it uses earnings, either alone or as part of a combination plan. This issue is discussed further in the next chapter.

UNIONS

A related issue is whether unionized employees should be members of the profit-sharing plan. Chapter 17 provides a brief review of the area of unions and profit sharing or any other form of variable pay.

It is also important to briefly review the legal framework for labour legislation in Canada. Unless a company is in a federally regulated industry such as banking, communications or railroads, it will be subject to the labour laws of the province(s) in which it operates. That is, if a company has operations in British Columbia and Manitoba and is not a federally regulated company, the employer should refer to the labour laws of both provinces.

As long as profit sharing is permitted (or not prohibited) by the relevant labour legislation or the current collective agreements and the union leadership is agreeable, there is no technical reason why union members cannot be members of the profit-sharing plan. However, it is important to ensure that all members of the plan are subject to the same rules and regulations. Although this may require delicate negotiations, do not set up separate rules for union and non-union employees.

Three of the companies whose plans are outlined in Chapter 19, "Sample Profit-Sharing Plans," have union members in their profit-sharing plan. They are Valley City Manufacturing, Algoma Steel, and Fisheries Products International.

OTHER CONSIDERATIONS

Owners of Canadian-controlled private corporations (CCPCs) and their relatives are prohibited from being members of a deferred profit-sharing plan (section 204, *Income Tax Act*). This restriction only poses a problem when a large number of family members are involved in the company. If the corporation is owned by just one person, it is not difficult to have him or her outside the plan. However, if a large number of family members (owners) are involved throughout the company at various levels, you could have a situation where employees (non-family) and family members are working side by side with one in the plan and the other outside it. Since one of the benefits of profit-sharing plans is that they create a commonality of interest between owners and employees, this can be problematic. In such a case, employers often use an employees profit-sharing plan instead of a DPSP.

7
ALLOCATION: DIVIDING THE FUNDS

This chapter explains the final "major"[1] issue that is part of all types of profit-sharing plans, cash and non-cash: allocation of the funds.

The "allocation" formula is simply the way in which monies placed in the profit-sharing fund by the company will be divided among the plan members. It is one of the most important decisions to be made in the design of a profit-sharing plan and will probably have more effect on the employees' perception of fairness and connection with productivity than any other feature of your plan.

The profit-sharing fund can be divided in a number of ways. Your choice of objective, which was discussed in Chapter 6, will have a significant impact on the formula. The following is a list of factors most often used for allocation purposes. They can be used alone or in various combinations:

- employee earnings
- job levels

[1] The two other major issues are the size and nature of the company contribution and membership in the plan.

- seniority or length of service
- attendance
- contributions by the employee
- merit or individual performance ratings
- equal distribution

Each factor is described in the sections following and examples using a particular employee are provided. There is also a section for three of the most common combinations: length of service/earnings; length of service/job level; and merit ratings/earnings. The examples all include the following assumptions:

- the employee is an engineer with 11 years of service and a base salary of $32,000 per year. Her name is Lorrie Pierce;
- the profit-sharing pool is $180,000;
- the total cash payroll is $4,528,000 per year; and
- the total base salaries are $4,066,000.

Chapter 8 provides further examples for each of the sections, which allow you to test various combinations of the employer contribution formula, membership, and allocation. These comprehensive formulas are also contained on the computer disk that is included with this book. Use of the disk is described in detail in Chapter 8.

EARNINGS

Earnings refer to the individual earnings of each employee. It is important to clearly define these earnings; they can be the base salary only or total earnings including overtime, shift and other premiums, and bonuses or commissions. It is critical that the definition of "earnings" be the same for all employees. If some employees are eligible for overtime but others are not, it is probably fairer to use base earnings. On the other hand, if every member of the plan has the same opportunity to earn overtime pay (or shift premiums or commissions), it is administratively easier to use total earnings. If your fiscal year ends on December 31, you can refer to total earnings on T4 slips prepared for income tax purposes.

The basic rationale for using earnings is that an employee's pay will typically reflect his or her relative contribution to the overall success of the corporation. This normally assumes that the employer has a rational, internally equitable (see Chapter 4 for a discussion and definition of "internal equity") and defensible system for determining base pay; if this is not the case, earnings should not normally be used in the allocation formula.

An alternative to using the actual base salary (or base plus other earnings) of individual employees is to use the mid-point of the salary range for the job of the employee. This approach has the advantage that it eliminates the problem described above regarding earnings other than base salaries. A further advantage is that it treats all employees in the same job level (e.g., salary grade X) similarly. Since salary ranges are often very wide, more senior employees can be very high in the range (they could be paid as much as 50 per cent or more than staff at the bottom of the same range). You are, in effect, recognizing length of service by using actual base salaries. If you also use length of service as part of a combination formula, this approach can double the effect of seniority on the payouts to individuals. Using salary range mid-points also requires that the employer has a formal salary administration system; if such a system is not in place, this is not an option.

The question of whether to include employees on commission (usually sales staff), bonus or productivity gain-sharing plans in the profit-sharing plan was discussed in Chapter 6. If you have decided to make them members of the plan, it is necessary to determine which definition of their earnings will be used. If you have decided, at least tentatively, to use base salaries, it is probably unfair to use them for the salespeople as well since their base pay will often be a small proportion of their total earnings. Their base salary will not reflect their relative contribution to profitability in the same way as it does for other, non-commissioned employees. The critical point to examine is the ratio of base salary to total earnings for the "average" sales representative. As a rule of thumb, I suggest that if base pay comprises 85 per cent or more of total earnings, it is reasonably fair to use base pay. If it does not, I suggest that earnings for sales employees be defined as base pay (by individual) plus the average (or median) commission for all sales staff.

EXAMPLE

The formula for using *base* earnings by itself is as follows:

$$\frac{\text{Employee salary}}{\text{Total salaries}} \times \text{Profit-sharing fund} =$$

$$\frac{32,000}{4,066,000} \times \$180,000 = \$1,416.63$$

Lorrie's share of the profit-sharing fund will be $1,416.63. This represents 4.43 per cent of her base salary.

JOB LEVELS

The use of job levels is based on the same rationale as using earnings; job level reflects relative contribution to profitability. However, it does so with less precision than earnings. It is a useful approach when you favour the rationale for using employee earnings but do not have the underlying salary administration system described earlier.

Job levels work as follows: all jobs are divided into groups such as senior management, middle management, first-line supervisors and professional staff, and all non-supervisory employees. Certain portions of the fund can be assigned to each group and then those monies are divided according to other criteria. Disadvantages of job levels are that they create "different classes of citizens" and may detract from the spirit of common purpose that profit-sharing plans are supposed to create. In addition, job levels cannot be used alone; they must be combined with some other factor such as earnings or length of service.

EXAMPLE

All employees could be grouped into one of the following four groups and the percentage shown taken from the profit-sharing pool and divided among the members of each group according to some other criteria. In this example the funds are divided equally among the employees in the job level or group.

GROUP	PERCENTAGE OF POOL	AMOUNT TO BE DIVIDED
A Top management	10% × $180,000	$18,000
B Middle management	25% × $180,000	$45,000
C First-level supervisors	15% × $180,000	$27,000
D All others	50% × $180,000	$90,000
	TOTAL 100%	$180,000

If the top management group consisted of three employees (a president and two vice presidents) and, assuming it was divided equally, each person would receive $6,000 according to the following formula:

$$\frac{\text{Percentage of fund for Group A} \times \text{Fund}}{\text{Number of employees in Group A}} =$$

$$\frac{10\% \times \$180,000}{3} = \$6,000$$

SENIORITY OR LENGTH OF SERVICE

The rationale for using seniority or length of service is that employees who have longer service with the company have greater influence on profitability than relatively newer employees. First, they have greater knowledge and are better at their jobs simply because they have more experience. Second, actions they took in previous years may have an influence on the current year's results and they should be rewarded for this—thus encouraging them to take a long-range perspective.

A major issue related to length of service is whether it should be capped. That is, should an employee continue to acquire extra points in the allocation formula without any maximum? In other words, does an employee's contribution to profits continually increase, assuming that she stays in the same job, or jobs at the same level, for the next 30 or 40 years, or does the rate of increase in contribution "top out" after some period of time, such as 10 years? Unless the employee changes jobs, it is unlikely that her value to the organization continues to increase indefinitely.

Consequently, many organizations cap seniority at around 10 years. However, you must also consider your objectives in establishing a plan; if the major focus is to reduce turnover, for example, it may be counter-productive to cap length of service.

If part-timers are included in the profit-sharing plan, it is necessary to specify how their service will be calculated. It is easiest and fairest to use the same method as that used in calculating the waiting period to become members of the plan. Refer to Chapter 6 for further details.

You must also decide how to calculate the number of years of service. Normal practice is to count the years at the end of the fiscal period (year, quarter, etc.) under review, but to only consider full years rather than rounding to the nearest year (which is administratively easier).

For example, an employee who started on October 1, 1991, will have four years as of October 1, 1995, and also as of December 31 if that is the fiscal year-end. If you used fractions of years, this employee will have 4.25 years.

The final issue regarding length of service is deciding at which point to begin calculating length of service. In other words, do you recognize service that was acquired before the introduction of the profit-sharing plan or only service since the plan was started? I suggest that it is fairer to recognize all service, but, again, this decision should reflect the culture of your organization.

EXAMPLE

Three points are given for each year of service and there is no cap or maximum. Assume that the total service points for the 150 employees, including Lorrie, are 2328. The formula is

$$\frac{\text{Employee points}}{\text{Total points}} \times \text{Profit pool} =$$

$$\frac{\text{Years of service} \times 3}{\text{Total years} \times 3} \times \$180,000 =$$

$$\frac{11 \times 3}{776 \times 3} = \frac{33}{2328} \times \$180,000 = \$2551.55$$

Lorrie will receive $2551.55.

ATTENDANCE

Attendance can also be used for allocating the profit-sharing fund. This is most effective where one of the objectives of the plan is to reduce absenteeism. Actual hours worked are counted and translated into shares of the fund.

However, you have to consider more than just hours worked. Items such as vacation hours taken should also be counted, otherwise the plan will discriminate against employees with longer service who are entitled to more vacation. Usually all absences, except for vacation, are not counted (since they are not hours worked) although you may also want to apply the same principle (as used for vacation) to other absences that are not within the employee's control, such as bereavement leave or jury duty.

Another area that may prove problematic is employees who "telecommute," that is, they work at home at least some of the time. In these cases, the employer is unable to physically or visually confirm that the employee is working. In this situation, it is probably most effective to use standard working hours (e.g., seven per day) as the basis for the calculation or not use attendance at all since it is largely a non-issue under such an arrangement.

The problem that overtime hours create when using earnings (as described above) also occurs when you use actual hours worked as all or part of the allocation formula. The same solution also applies. If some employees work overtime that is recorded and some do not, it is probably best to use "normal" hours worked and exclude overtime hours as well as other hours worked such as "call-in" or "stand-by" hours.

EXAMPLE

Only standard (i.e., other than overtime, call-in, or emergency work) hours worked will be counted. All staff work 40 hours per week or 2080 hours per year. The average employee has lost 20 hours in the past year to various absences. Lorrie, our sample employee, has worked 2075 hours. Her share of the profits will be

$$\frac{\text{Hours worked}}{\text{Total hours worked}} \times \$180,000 =$$

$$\frac{2{,}075}{(2{,}080 - 20) \times 150} \times \$180{,}000 =$$

$$\frac{2{,}075}{309{,}000} \times \$180{,}000 = \$1{,}208.74$$

EMPLOYEE CONTRIBUTIONS

Until the *Income Tax Act* was changed in 1991, employees were allowed to make contributions (subject to specified limits) with their own funds, to deferred profit-sharing plans. Although these contributions were not tax-deductible for the employees, the earnings and capital gains they accrued were sheltered from income tax in the same way that contributions by the employer corporation were. Such employee contributions are no longer permitted.

When they were permitted, some employers made employee contributions mandatory with the same objective as making contributions to a registered pension plan, i.e., to encourage employees to save for their retirement. Other employers made contributions voluntary but encouraged them by basing at least part of the allocation formula on the size of the employer contribution.

Now, some employers continue this tradition by setting up a group RRSP in addition to the DPSP (see Dun & Bradstreet in Chapter 19, "Sample Profit-Sharing Plans," p.195) and either wholly or partially matching the payments by the employee to the RRSP. The payments by the employer go into the deferred plan. This can also be done with an EPSP.

The advantage of this approach is that it can encourage employee savings for retirement or other future goals. A disadvantage is that it may unfairly reward higher-paid employees, who will have more discretionary income to place in tax shelters. There is also the problem that employees could place monies in the RRSP and, after the employer payments are vested, remove the funds from the RRSP. Remember that the employer cannot control the funds in the RRSP; the employee can remove them at any time provided that he or she is willing to pay the tax owing.

EXAMPLE

The employer has agreed to pay 50¢ for every $1.00 contributed by the employee. Lorrie believes that she can afford a contribution of $1,500.00 to the RRSP.

Her share of the profits will be
$$\$1{,}500.00 \times \$0.50 = \$750.00$$

The employer will pay $750.00 into the DPSP in Lorrie's name.

MERIT RATINGS

Some plans based the allocation, at least in part, on the performance rating of the individual (approximately 9 per cent of plans according to the Hewitt Survey[2]). I do not recommend this for two reasons.

First, it is contrary to the overall theory of profit sharing, which holds that the effectiveness of these plans is due to the fact that they encourage group cohesiveness and teamwork whereas individual pay plans do not. As an employer, if you believe that it is important to have some form of individual "pay for performance" component, it is better to keep it out of the profit-sharing plan and put it in the base salary. See Chapter 16 for further discussion of this topic.

The second reason is that the requirements for a workable merit component in the allocation formula are quite stringent. The most obvious is the need for a performance management system. If you don't already have one, designing and implementing such a system is a major project in itself and can take up considerably more time and effort on the part of management than the profit-sharing plan itself. Since it requires specialized expertise, it is usually necessary to hire outside consultants to assist in this process. Extensive training of individual employees, managers and supervisors is also required. However, a more significant prerequisite for a performance management system is a high level of trust in management by the employees and a willingness on the part of all those with supervisory responsibilities to conduct meaningful assessments of the performance of their staff. These are difficult requirements; they are almost impossible to meet during the design of a profit-sharing plan unless they are already in place.

EXAMPLE

The company has a performance management system with five levels of performance defined and points given for each level as follows:

Rating	Points
Outstanding	15
Good	10

[2] Hewitt Associates, *Survey of Canadian Profit-Sharing Plans* (North York, Ontario, 1990).

Rating	Points
Meets job requirements	5
Needs improvement*	2
Completely unsatisfactory	0

Lorrie's performance has been reviewed by her manager and been rated at the "Outstanding" level. The 150 employees in the company (including Lorrie) have received a total of 797 merit points.

Her share of the profits will be

$$\frac{\text{Her performance points}}{\text{Total performance points}} \times \$180,000 =$$

$$\frac{15}{797} \times \$180,000 = \$3,387.70$$

*You are probably wondering why an employee who "needs improvement" should get any points. The theory is that most employees with this rating are probably capable of raising their performance to the next level and to cut them out entirely puts them in the same category as the "completely unsatisfactory" employee who is unlikely to improve.

EQUAL DISTRIBUTION

About 9 per cent of plans provide that the profits in the fund be split equally among the members of the plan (In 1992 the percentages were by type of plan: cash 14%; DPSP 4%; EPSP 0%, and combination 1%.)[3] This is usually done in smaller organizations with a philosophy that "we are all in this together and we are all trying as hard as we can." Equal allocation is also appropriate where the organization structure is relatively flat and there is not much difference between the highest and lowest salaries.

EXAMPLE

Each of the 150 employees in the company will receive

$$\frac{\text{Profit pool}}{\text{\# of employees}} = \frac{\$180,000}{150} = \$1,200$$

For Lorrie, this payment represents 3.75 per cent of her annual salary.**

**This test, expressing the payment as a percentage of the employee's salary for the fiscal period, should be done when testing any formula.

[3] Hewitt Survey, 1990, 11, 17, 40, 51.

COMBINATIONS

Each of the usual factors used in allocation formulas has already been discussed separately above. However, few plans in Canada use one factor by itself. Most plans use two or more in combination.

Combination formulas allow the plan to meet a variety of objectives and this is the primary way to choose the factors that you wish to use in your allocation formula. Also consider your original objective(s).

It is also important to determine the relative importance of the two or more criteria that you, or the committee, choose to use. They do not have to be equal in their impact on allocation; it is usually possible to construct the formula so that the criteria used reflect their importance relative to your initial objectives.

Some examples of the most common combinations of factors are shown and discussed below. All of them use the same assumptions as the single factor examples shown above. The rationale for using the factors is not reviewed since each has been discussed earlier.

LENGTH OF SERVICE AND EARNINGS

Employees are given two points per year of service to a maximum of 10 points (i.e., five years) and one point per $1,000 of base salary. Lorrie will have total points of

$$5(\text{years}) \times 2 = 10$$
$$32('000\text{'s}) \times 1 = 32$$
$$\text{Total} \quad 42$$

If the total points for all members of the plan are 5,372, this employee's share will be

$$\frac{\text{Service points + Salary points}}{\text{Total points}} \times \text{Profit pool} =$$

$$\frac{10 + 32}{5,372} \times \$180,000 = \$1,407.29$$

If you felt that salary should have more influence in the formula, you could change it so that each $1,000 of base salary received five points. The total for all members then becomes 21,636 (5372 + (4 × 4066)), and Lorrie's total will be

$$5(\text{years}) \times 2 = 10$$
$$32('000\text{'s}) \times 5 = 160$$
$$\text{Total} = 170$$

Her share will be

$$\frac{170}{21,636} \times \$180,000 = \$1,414.30$$

As you can see, this change in points has very little impact on Lorrie's share of profits because her salary is one of the lowest in this group. However, if you do the same calculation for the general manager, William Dixon, his share goes from $5,361 to $6,323.

LENGTH OF SERVICE AND JOB LEVEL

As in the example given above an employee receives two points for each year of service but to a maximum of 10 years. All jobs are divided into four groups: A-executive, B-middle management, C-supervisors and non-managerial professional/technical employees, and, D-all others. Lorrie is an engineer in group C and has 20 seniority points. Seniority points for this C group total 40. The profit-sharing pool is divided among the four groups as follows:

A - 10%
B - 25%
C - 15%
D - 50%

Lorrie would calculate her share in this way:

$$15\% \times \$180,000 \times 20/40 = \$13,500$$

This amount is 42.19 per cent of Lorrie's base salary and would be considered excessive by most since it is by far the highest payout, expressed as a percentage of base, of any of the employees. If you wished job level to have less influence on group C, the distribution between job levels could be changed to:

A - 10%
B - 20%
C - 5%
D - 65%

In this case her share would be

$$5\% \times \$180,000 \times 20/40 = \$4,500$$

Merit Ratings and Earnings

The company uses a merit rating system with a five-point scale. Points are given for different levels of performance as follows:

Rating	Points
Outstanding	15
Good	10
Satisfactory	5
Needs improvement	2
Completely unsatisfactory	0

Two points are awarded for each $1,000 of base earnings. Lorrie has been rated as "Outstanding" by her supervisor and manager. Total points for all employees are 9,004. Her point total is:

$$\text{Earnings points} = 32 \times 2 = 64$$
$$\text{Merit points} = 15$$
$$\text{Total points} = 79$$

Her share of profits will be = $\dfrac{79}{9,004} \times \$180,000 = \$1,579.28$

If you wished to make the merit rating much more influential, the points could be

Rating	Points
Outstanding	80
Good	40
Satisfactory	5
Needs improvement	2
Completely unsatisfactory	0

In this situation her point total will be

$$\text{Earnings points} = 2 \times 32 = 64$$
$$\text{Merit points} = 80$$
$$\text{Total points} = 144$$

Her share will be $\dfrac{144}{9769} \times \$180,000 = \$2,653.29$

HOW TO CHOOSE AN ALLOCATION FORMULA

Given all the choices involved, how do you choose an allocation formula for your profit-sharing plan? You can use three major criteria:

- ease of communication
- degree of acceptability to the members of the plan
- relationship to the objective(s) of the plan

OBJECTIVE(S) OF THE PLAN

The most important criterion is relating the allocation formula to the original objective or objectives that you established when you started this project. The allocation formula must reinforce the objective(s). For example, if one objective was to reduce turnover, the formula will have a strong seniority component. If, however, you wish to encourage improvement in the performance of individual employees, the formula must include merit rating. Use of earnings or job levels will encourage employees to seek advancement to higher-level or higher-paid jobs. If the objective includes improving teamwork, the formula will be structured so that payouts do not vary widely and will not include an individual merit component.

EASE OF COMMUNICATION

The second criterion is the degree to which the plan is easy to explain to the employees. There is a common tendency to make allocation formulas overly complex. Complexity permits you to have a formula that is technically perfect, but such formulas are always difficult to communicate to the average employee. Simplicity is a major virtue in designing an allocation formula and it also simplifies administration.

ACCEPTABILITY

The last way to assess the proposed formula is highly subjective but very important. Is it acceptable to the members of the plan?

Do they think that it is fair? This assessment can best be done through your employee committee and is one of the best arguments for using an employee committee. If you have developed the plan using the procedures outlined in Chapter 5, this test should be straightforward. However, as a final wrap-up you should ask the committee: will employees think that this formula is a reasonable way to divide the profit-sharing fund? How do you gauge employee opinion if you have decided against forming an employee profit-sharing committee? One way is to just make a judgement call based on your knowledge of the employees and the company culture. However, for the reasons discussed above in Chapter 5, this approach is highly unreliable. Another option is to put the entire plan, including the allocation formula, to a vote. This is a high-risk option; if you do not get the favourable vote you hoped for, you will be in a position where you have to change the plan or scrap it entirely. For this reason, I highly recommend the use of an employee committee.

8
TESTING YOUR FORMULA

This chapter reviews the three major components of a profit-sharing plan — the amount placed in the fund, the membership rules and the allocation formula — and shows you how to put them together so that they can be tested before implementation.

Once the profit-sharing committee has had its second meeting (assuming you have chosen to involve employees), you will have made tentative decisions about

- the definition of profits and the percentage of company profits to be placed in the profit-sharing fund;
- who will be members of the plan; and
- the factor(s) to be used in the allocation formula.

The first two items are relatively constant and can be easily tested at this point. However, if the allocation formula is a combination of factors — and most are — you must still decide how they will be combined. Even if the allocation formula uses only one factor, this is still an appropriate time to test the overall impact of the tentative decisions made about the three variables. Testing at this stage is important for the following reasons:

- it allows you to see the impact of various combinations;
- it illustrates the payments, and, therefore, the effect on the individual;
- it facilitates the use of actual, historical numbers so that you can see what would have happened under actual circumstances; and
- it allows you to test the future effects of the formula using your strategic plan or financial forecasts.

Testing the formula is essential and I recommend that you do not proceed until this has been done. You can develop your own model using a spreadsheet program that you are familiar with or you can use the ten models provided, at the end of this chapter.

They are organized according to the allocation formulas described in Chapter 7 because the allocation formula is the most variable of the three main issues involved. The other two are the definition of membership and the percentage of profits allocated to the profit-sharing pool. All sample worksheets allow you to vary profit levels and the percentage of profit allocated to the plan. All worksheets require you to list the employees, either individually or by group.

These worksheets, with the appropriate formulas, are also contained on the floppy disk included with this book. If you decide to test any one of the options presented, you can do so using this disk. If you decide to use a formula that is not included in this book, you may be able to adapt one of the standard formulas to your circumstances.

The formulas were written using the spreadsheet program, Quattro Pro®, Version 5.0. Most, if not all, of the popular spreadsheet programs can read Quattro Pro® files. Consult the manual for your program. The disk enclosed includes these spreadsheets not only in Quattro Pro®, but also in Microsoft® Excel and Lotus® 123®.

Each option is built around variations on the allocation formula and allows you to test various combinations of

- profit levels;
- percentage of profit to be placed in the fund;
- membership; and
- the allocation formula.

You can use historical, current and/or forecasted numbers for each variable. Testing can be done regardless of whether the plan is a cash, DPSP, EPSP or combination plan.

In using these model formulas there are a number of general guidelines that apply to them all. These are discussed below. Rules specific to each model are provided in the section for that model which are as follows:

OPTION	FILE NAME	FACTOR(S) IN ALLOCATION FORMULA	PAGE CHAP 8	PAGE CHAP 7
A	A1.WQ2	Earnings	82	58
B	B1.WQ2	Job Levels	83	60
C	C1.WQ2	Seniority/Length of Service	84	61
D	D1.WQ2	Attendance	85	63
E	E1.WQ2	Employee Contributions	86	64
F	F1.WQ2	Merit Ratings	87	65
G	G1.WQ2	Equal Distribution	88	66
H	H1.WQ2	Length of Service/Earnings	89	67
I	I1.WQ2	Length of Service/Job Level	90	68
J	J1.WQ2	Merit Ratings/Earnings	91	69

The Options listed above are discussed in this chapter and the actual results of all the formulas and calculations are shown in the spreadsheets at the chapter's end. The underlying formulas are shown in Appendix 2 in the same format except that, in several cases, the columns have been widened in order to show the complete formula.

ASSUMPTIONS

The sample formulas in this chapter use the same assumptions as the examples in Chapter 7 as well as some additional ones.

ASSUMPTIONS FROM CHAPTER 7

- the company has 150 employees;
- total annual cash payroll is $4,528,000;
- profit-sharing pool is $180,000;
- total base salaries are $4,066,000;

- one employee, Lorrie Pierce, is an engineer with 11 years of service and an annual salary of $32,000.

ADDITIONAL ASSUMPTIONS

- The company has a relatively small group of managerial/professional employees, including Lorrie, the engineer. They are as follows:

NAME	TITLE	ANNUAL SALARY	OTHER EARNINGS*
William Dixon	President	150,000	45,000
Lynda Harris	Vice President	100,000	22,000
Robert Campbell	Vice President	95,000	15,000
Jeannette Totten	Accounting Manager	70,000	—
Ann Walker	H. R. Manager	65,000	—
Neil Prior	Production Manager	75,000	—
Jack Kelly	Purchasing Manager	68,000	—
James Curtis	Sales Manager	72,000	—
Nancy Dawson	Customer Service Mgr.	48,000	—
Lorrie Pierce	Engineer	32,000	—
Susanna Jackson	Senior Sales Rep.	25,000	15,500
Thomas Whicher	Accountant	32,000	—
**various 12	Sales Representatives	20,000	10,000
various 15	Salaried Non-Supervisory	22,000	1,500
various 111	Hourly, Production & Warehouse	24,000	2,000

* "Other Earnings" can include a number of items such as bonuses, overtime, shift premiums or call-in pay. This item only includes payments in cash. In your own testing you may wish to breakdown "other earnings" into the component parts. Refer to the section in Chapter 7 for a review of the definition of earnings.

** The numbers below this line refer to the averages (both base salary and other earnings) for the respective groups listed. Averages are used for the sake of simplicity in showing the formula. You may wish to list every employee when doing your own test.

- Only full-time employees are included in the plan.

GENERAL RULES

These are covered according to the standard headings that are used in all options. To simplify matters, all cell references are to Option A. They are usually the same in the other options but may vary and readers are cautioned to be aware of this.

FISCAL PERIOD
Fill in the fiscal period (year, quarter, etc.).

PROFITS
Fill in the amount of profits for the fiscal period using the definition of profits that you have selected (see Chapter 5).

PERCENTAGE OF PROFITS
Fill in the percentage of profits that you are planning to use.

COLUMN A — SALARY GRADE/JOB LEVEL[1]
If you do not have a formal salary administration system, you will not have salary grades. If this is the case, you can create your own set of job levels by listing them from highest to lowest as in the example for Option B1. Otherwise, leave this column blank and list the employees from highest to lowest base salary in Column C.

COLUMN B — EMPLOYEE NAME
Fill in the name of the employee. Although it is not necessary, you may choose to input the job title as well. This can be done either by widening Column B or by inserting a new Column C. If you have multi-incumbent job classes, you can save inputting time and space on the spreadsheet by grouping employees and not inputting each individual's name. Place the name of the job class in Column A. However, you must also input the number of employees in the job class in Column B. See any of the Options included in this chapter as examples.

COLUMN C
This is the base salary stated in annual terms in descending order of magnitude. It is by individual except for the multi-incumbent

---

[1] Columns are lettered above the box with the thick black lines. Naturally, the column extends both up and down. Row numbers are provided along the right side of the spreadsheet.

job classes (Cells C23-25 in Option A) in which you must insert the average salary of the employees in the job class. The formula in those cells then multiplies this salary by the number of employees to arrive at the total base salaries for the job class.

COLUMN D — OTHER EARNINGS

Input cash earnings (other than salary) such as bonuses, commissions, overtime and premiums of all sorts according to the definition of earnings you have developed (see Chapter 7, page 58). This column will not necessarily be in order of magnitude as is Column C. This column is like Column C in that it includes both individuals and job classes.

Note that unless you are using earnings, either base or total, as all or part of the allocation formula, you do not have to complete Columns C and D. However, the data are extremely useful in applying the test of the percentage of the employee's pay (either base or total) represented by his or her share of profits. I highly recommend doing so. This is one of the most effective ways to evaluate the "fairness" of the allocation formula. Columns G (Payout as a percentage of base salary) and H (Payout as a percentage of total earnings) cannot be computed unless you complete Columns C and D. For example, in Option A, the calculation in Column G shows that all employees will receive 4.43 per cent of base salary in profits, which meets the test that any payout in profits should be 3 to 5 per cent of the employee's pay. Column H gives the same result, although one employee, Susanna Jackson, and the Sales group, are slightly below the 3 per cent target.

COLUMN E — TOTAL EARNINGS

No input is necessary as the program adds Columns C and D.

COLUMN F — PAYOUT

This is the actual dollar amount of profit sharing to be paid to the employee. No input is necessary.

COLUMN G — PAYOUT AS PERCENT OF BASE SALARY

This column shows the payout from Column F as a percentage of the base salary. No input is necessary.

COLUMN H — PAYOUT AS PERCENT OF TOTAL EARNINGS

This column shows the payout from Column F as a percentage of the total earnings in Column E. No input is necessary.

PAYOUT PER PERSON FOR MULTI-INCUMBENT JOB CLASSES

This separate box in the lower left-hand corner of the spreadsheet shows the actual dollar amount to be paid to each individual in job classes with more than one incumbent. In the examples, these job classes are Hourly, Salaried and Sales employees.

TESTS OF REASONABLENESS

The spreadsheets in the examples provided have been designed with several quick and easy "tests of reasonableness" that enable you to quickly assess whether the formulas are working correctly. These are especially important if you decide to customize the formulas provided. These tests should also be built into any of your own spreadsheets if you decide to prepare one from scratch. The tests provided in this section are common to all the examples. Some other tests are provided that are unique to that option and are listed below the sample spreadsheet for each option. The common tests are

1. Cell F26 (total of Column F) should be equal to Cell B7 (Fund).
2. Cell E27 (the sum of the totals of Columns C and D) should be equal to Cell E26 (the sum of Column E).
3. Column H (the payout as a percent of Total Earnings) should be less than the results in Column G (the payout as a percent of Base Salary) unless the entry in Column D — Other Earnings — is 0, in which case the two percentages should be the same.
4. The percentage shown at the bottom of Column H should always be less than the percentage at the bottom of Column G unless Column D (Other Earnings) is all zeros.

TESTING

Follow these steps when you decide to test your formulas.

1. Select the Option you prefer or develop your own spreadsheet model.

2. Input the amount of profits from the last fiscal period.
3. Input the percentage of profits that you think is appropriate having followed the process described in Chapter 5.
4. Check the payouts to individuals and jobs classes in terms of the actual dollar amounts and the percentages of base salary and total earnings. Look for essential fairness and the three to five per cent rule (see Chapter 5; page 46).
5. If the test fails, repeat steps 3 and 4.
6. Start altering the variables in whichever option you have chosen.
7. Once you have developed a tentative formula for the most recent fiscal period, test it with profit levels from the past three to five years and whatever projected numbers you have.
8. Finalize the overall formula.

COMMUNICATIONS

The use of the spreadsheets is an essential part of testing variations of your proposed plan. However, this creates a problem in terms of communications and the operation of your employee profit-sharing committee.

The issue relates to confidentiality. In terms of receiving feedback about the perceived fairness of the proposed formula that you have chosen after manipulating the spreadsheet, you will probably want to show the results of your calculations to the committee. This would mean, however, that if you used the formats provided in this chapter, you would be showing salaries, other earnings, and total earnings, by individual, or groups, to the committee. Most employers are extremely reluctant to do this. Even if the committee members are sworn to secrecy, some of the information is almost certain to leak out. Another concern is that if you do not have a fair system of cash compensation, committee members may focus on the system's inequities (refer to Chapter 4, especially the sections on internal equity and external competitiveness) and lose sight of the potential merits of the profit-sharing plan.

Another way to disclose the test results shown in the format in this chapter is to group the employees listed individually by name in the same manner as the Hourly, Salaried and Sales Staff have been grouped. A problem with this approach is that averages may be meaningless because of the wide range of earnings. For example, for the 12 employees listed in our sample on page 76, base salaries range from $25,000 to $150,000. This problem can be alleviated by making the grouping smaller, for example, using job levels as in Option B1. This can be done even if Job Levels is not part of the allocation formula.

Another option is to show the committee only the bottom line of the box with the thick black lines.

The decision on how much information to disclose to the committee is a highly subjective one. To make this decision, consider the following issues: your company's past history of disclosure; the underlying fairness of your existing salaries and other earnings; and the level of trust in management by the employees.

Profit-Sharing Plan Worksheet
Option A1 Allocation = Base Earnings Only

	A	B	C	D	E	F	G	H
Fiscal Period	94-95							
Profits	$1,500,000							
% of Profits	12							
Fund	$180,000							
Salary Grade/ Job Level	Name of Employee	Base Annual Salary	Other Earnings	Total Earnings	Payout	Payout as % of Base	Payout as % of Total
	William Dixon	$150,000	$45,000	$195,000	$6,640	4.43%	3.41%
	Lynda Harris	$100,000	$22,000	$122,000	$4,427	4.43%	3.63%
	Robert Campbell	$95,000	$15,000	$110,000	$4,206	4.43%	3.82%
	Neil Prior	$75,000	$0	$75,000	$3,320	4.43%	4.43%
	James Curtis	$72,000	$0	$72,000	$3,187	4.43%	4.43%
	Jeannette Totten	$70,000	$0	$70,000	$3,099	4.43%	4.43%
	Jack Kelly	$68,000	$0	$68,000	$3,010	4.43%	4.43%
	Ann Walker	$65,000	$0	$65,000	$2,878	4.43%	4.43%
	Nancy Dawson	$48,000	$0	$48,000	$2,125	4.43%	4.43%
	Thomas Whicher	$32,000	$0	$32,000	$1,417	4.43%	4.43%
	Lorrie Pierce	$32,000	$0	$32,000	$1,417	4.43%	4.43%
	Susanna Jackson	$25,000	$15,500	$40,500	$1,107	4.43%	2.73%
Hourly	111	$2,664,000	$222,000	$2,886,000	$117,934	4.43%	4.09%
Salaried	15	$330,000	$22,500	$352,500	$14,609	4.43%	4.14%
Sales	12	$240,000	$120,000	$360,000	$10,625	4.43%	2.95%
		$4,066,000	$462,000	$4,528,000	$180,000	4.43%	3.98%
				$4,528,000			

Payout Per Person for Multi-Incumbent Job Classes

Hourly = $1,062
Salaried = $974
Sales = $885

Input

1. The only input required for this option is Cell B5 (Profits) and Cell B6 (% of Profits) and the fiscal period in Cell B4.

Tests

1. Column G — the percentages should all be the same.
2. Column H — the percentages should all be the same for all those employees whose Other Earnings are 0 (Column D).

Profit-Sharing Plan Worksheet

Option B1 Allocation = Job Levels/Equal Within Levels

Job Levels	%age of Fund	Number of Employees in Level
Top Management = A	10	3
Middle Management = B	25	6
First Level Supervisors = C	15	3
All others = D	50	138
Total	**100**	**150**

Category of Employee	Number
Hourly	111
Salaried	15
Sales Reps.	12

Fiscal Period: 94-95
Profits: $1,500,000
% of Profits: 12
Fund: $180,000

Salary Grade/ Job Level	Name of Employee	Base Annual Salary	Other Earnings	Total Earnings	Payout on Base	Payout as % of Base	Payout as % of Total
A	William Dixon	$150,000	$45,000	$195,000	$6,000	4.00%	3.08%
A	Lynda Harris	$100,000	$22,000	$122,000	$6,000	6.00%	4.92%
A	Robert Campbell	$95,000	$15,000	$110,000	$6,000	6.32%	5.45%
B	Neil Prior	$75,000	$0	$75,000	$7,500	10.00%	10.00%
B	James Curtis	$72,000	$0	$72,000	$7,500	10.42%	10.42%
B	Jeannette Totten	$70,000	$0	$70,000	$7,500	10.71%	10.71%
B	Jack Kelly	$68,000	$0	$68,000	$7,500	11.03%	11.03%
B	Ann Walker	$65,000	$0	$65,000	$7,500	11.54%	11.54%
B	Nancy Dawson	$48,000	$0	$48,000	$7,500	15.63%	15.63%
C	Thomas Whicher	$32,000	$0	$32,000	$9,000	28.13%	28.13%
C	Lorrie Pierce	$32,000	$0	$32,000	$9,000	28.13%	28.13%
C	Susanna Jackson	$25,000	$15,500	$40,500	$9,000	36.00%	22.22%
D-Hourly	111	$2,664,000	$222,000	$2,886,000	$72,391	2.72%	2.51%
D-Salaried	15	$330,000	$22,500	$352,500	$9,783	2.96%	2.78%
D-Sales Reps.	12	$240,000	$120,000	$360,000	$7,826	3.26%	2.17%
		$4,066,000	$462,000	$4,528,000	$180,000	4.43%	3.98%
				$4,528,000			

Payout Per Person for Multi-Incumbent Job Classes

Hourly =	$652
Salaried =	$652
Sales =	$652

Input
1. You can vary both of the columns "%age of fund" (E4-E7) and "Number of employees in level" (F4-F7).

Comments
1. The percentages (in Column G) seem excessively high for Job Levels B and C.

Tests
1. The payouts (Column F) within each Job Level should be equal to each other.
2. The "Payouts per person for multi-incumbent job classes" should all be the same.

Try adjusting them by changing cells E4-E7.

83

Profit-Sharing Plan Worksheet

Option C1 Allocation = Seniority/Length of Service

Assumptions
Points per year of service = 3   With no maximum

Fiscal Period	94-95
Profits	$1,500,000
% of Profits	12
Fund	$180,000

Salary Grade/Job Level	Name of Employee	Base Annual Salary	Other Earnings	Total Earnings	Payout on Base	Payout as % of Base	Payout as % of Total	Years of Service	Seniority Points	Row Number
	William Dixon	$150,000	$45,000	$195,000	$3,711	2.47%	1.90%	16	48	13
	Lynda Harris	$100,000	$22,000	$122,000	$5,103	5.10%	4.18%	22	66	14
	Robert Campbell	$95,000	$15,000	$110,000	$928	0.98%	0.84%	4	12	15
	Neil Prior	$75,000	$0	$75,000	$2,088	2.78%	2.78%	9	27	16
	James Curtis	$72,000	$0	$72,000	$2,784	3.87%	3.87%	12	36	17
	Jeannette Totten	$70,000	$0	$70,000	$464	0.66%	0.66%	2	6	18
	Jack Kelly	$68,000	$0	$68,000	$1,624	2.39%	2.39%	7	21	19
	Ann Walker	$65,000	$0	$65,000	$1,160	1.78%	1.78%	5	15	20
	Nancy Dawson	$48,000	$0	$48,000	$232	0.48%	0.48%	1	3	21
	Thomas Whicher	$32,000	$0	$32,000	$1,392	4.35%	4.35%	6	18	22
	Lorrie Pierce	$32,000	$0	$32,000	$2,552	7.97%	7.97%	11	33	23
	Susanna Jackson	$25,000	$15,500	$40,500	$1,392	5.57%	3.44%	6	18	24
Hourly	111	$2,664,000	$222,000	$2,886,000	$102,990	3.87%	3.57%	4	1332	25
Salaried	15	$330,000	$22,500	$352,500	$31,314	9.49%	8.88%	9	405	26
Sales Reps.	12	$240,000	$120,000	$360,000	$22,288	9.28%	6.19%	8	288	27
		$4,066,000	$462,000	$4,528,000	$180,000	4.43%	3.98%		2328	28
				$4,528,000						

Payout Per Person for Multi-Incumbent Job Classes

Hourly =	$928
Salaried =	$2,088
Sales =	$1,856

Input

1. Cells I25, 26, 27 are the average years of service for the respective groups of jobs, Hourly, Salaried and Sales.

2. For the rest of Column I input the actual years of service for each individual.

3. If you wish to have a maximum number of years you can either manually limit the input in Column I to that maximum OR place an @IF(Quattro Pro) command in Column J.

Comment

1. Note the difference in Payout as a Percent of Base between James Curtis and Lorrie Pierce whose years of service are almost identical (12/11) although the payout for Curtis is slightly higher. The reason is the difference in base salary. You need to examine if this is acceptable.

Profit-Sharing Plan Worksheet

Option D1 Allocation = Attendance (Regular Hours Only)

Assumptions

Average hours/year of non-management

Hourly =		2066
Salaried =		2071
Sales =		2075

Fiscal Period	94-95
Profits	$1,500,000
% of Profits	12
Fund	$180,000

	A	B	C	D	E	F	G	H	I
	Salary Grade/ Job Level	Name of Employee	Base Annual Salary	Other Earnings	Total Earnings	Payout	Payout as % of Base	Payout as % of Total	Hours Worked
		William Dixon	$150,000	$45,000	$195,000	$1,201	0.80%	0.62%	2061
		Lynda Harris	$100,000	$22,000	$122,000	$1,212	1.21%	0.99%	2080
		Robert Campbell	$95,000	$15,000	$110,000	$1,165	1.23%	1.06%	2000
		Neil Prior	$75,000	$0	$75,000	$1,191	1.59%	1.59%	2045
		James Curtis	$72,000	$0	$72,000	$1,206	1.67%	1.67%	2070
		Jeannette Totten	$70,000	$0	$70,000	$1,208	1.73%	1.73%	2073
		Jack Kelly	$68,000	$0	$68,000	$1,212	1.78%	1.78%	2080
		Ann Walker	$65,000	$0	$65,000	$583	0.90%	0.90%	1000
		Nancy Dawson	$48,000	$0	$48,000	$1,212	2.52%	2.52%	2080
		Thomas Whicher	$32,000	$0	$32,000	$1,203	3.76%	3.76%	2065
		Lorrie Pierce	$32,000	$0	$32,000	$1,209	3.78%	3.78%	2075
		Susanna Jackson	$25,000	$15,500	$40,500	$1,212	4.85%	2.99%	2080
	Hourly	111	$2,664,000	$222,000	$2,886,000	$133,588	5.01%	4.63%	229326
	Salaried	15	$330,000	$22,500	$352,500	$18,096	5.48%	5.13%	31065
	Sales Reps.	12	$240,000	$120,000	$360,000	$14,505	6.04%	4.03%	24900
			$4,066,000	$462,000	$4,528,000	$180,000	4.43%	3.98%	309000
					$4,528,000				

Payout Per Person for Multi-Incumbent Job Classes

Hourly =	$1,203
Salaried =	$1,206
Sales Reps. =	$1,209

Input

1. In Column I input the actual hours worked during the year by each individual named in Column B.

2. Input the average number of hours per year for each of the groups, Hourly, Salaried, and Sales in Cells D5, D6 and D7. Cells I27-I29 inclusive will show the total hours for each group.

Tests

1. If 2 or more employees have worked the same number of hours the $ payout in Column F should be the same, e.g., Harris, Kelly, and Jackson.

Profit-Sharing Plan Worksheet

Fiscal Period: 94-95
Profits: $1,500,000
% of Profits: 12
Fund: $180,000

Option E1 Allocation = Employee Contributions to Group R.R.S.P.

Assumptions
Average contribution $/year of non-management employee
- Hourly = $800
- Salaried = $500
- Sales = $3,000

Employer contribution(s) for each dollar of employee contribution = 0.5

	A	B	C	D	E	F	G	H	I	Row Number
	Salary Grade/Job Level	Name of Employee	Base Annual Salary	Other Earnings	Total Earnings	Payout	Payout as % of Base	Payout as % of Total	Employee Contribution	
		William Dixon	$150,000	$45,000	$195,000	$4,875	3.25%	2.50%	$9,750	14
		Lynda Harris	$100,000	$22,000	$122,000	$3,660	3.66%	3.00%	$7,320	15
		Robert Campbell	$95,000	$15,000	$110,000	$4,400	4.63%	4.00%	$8,800	16
		Neil Prior	$75,000	$0	$75,000	$1,125	1.50%	1.50%	$2,250	17
		James Curtis	$72,000	$0	$72,000	$2,160	3.00%	3.00%	$4,320	18
		Jeannette Totten	$70,000	$0	$70,000	$0	0.00%	0.00%	$0	19
		Jack Kelly	$68,000	$0	$68,000	$2,380	3.50%	3.50%	$4,760	20
		Ann Walker	$65,000	$0	$65,000	$250	0.38%	0.38%	$500	21
		Nancy Dawson	$48,000	$0	$48,000	$1,000	2.08%	2.08%	$2,000	22
		Thomas Whicher	$32,000	$0	$32,000	$500	1.56%	1.56%	$1,000	23
		Lorrie Pierce	$32,000	$0	$32,000	$750	2.34%	2.34%	$1,500	24
		Susanna Jackson	$25,000	$15,500	$40,500	$1,500	6.00%	3.70%	$3,000	25
	Hourly	111	$2,664,000	$222,000	$2,886,000	$44,400	1.67%	1.54%	$88,800	26
	Salaried	15	$330,000	$22,500	$352,500	$3,750	1.14%	1.06%	$7,500	27
	Sales Reps.	12	$240,000	$120,000	$360,000	$18,000	7.50%	5.00%	$36,000	28
			$4,066,000	$462,000	$4,528,000	$88,750	2.18%	1.96%	$177,500	29
					$4,528,000					30

Payout Per Person for Multi-Incumbent Job Classes
- Hourly = $400
- Salaried = $250
- Sales Reps. = $1,500

Input
1. Fill in the average contribution in $ per year of each of the groups, Hourly, Salaried and Sales in Cells G5, G6 and G7 respectively. The totals for each group will be calculated automatically in Cells I26-28.
2. In Column I fill in the individual contribution to his/her RRSP in Cells I14-25.
3. Input the employer contribution in $ for each dollar of employee

Notes
1. This is the only option in which the total of Column F (Cell F29) does not equal the size of the Fund (Cell B7). This is because the payouts are not governed by the size of the fund but by the employees contributions to the RRSP. However, you could use the Fund (B7) as an indicator of the maximum to be paid and adjust Cell H10 until the total of Column F is just under the amount of the fund.

Tests

Profit-Sharing Plan Worksheet

Option F1 Allocation = Merit Ratings

Assumptions

Merit rating system provides points as follows:

- Outstanding = A = 15
- Good = B = 10
- Meets job requirements = C = 5
- Needs improvement = D = 2
- Completely unsatisfactory = E = 0

Fiscal Period	94-95
Profits	$1,500,000
% of Profits	12
Fund	$180,000

Salary Grade/Job Level	Name of Employee	Base Annual Salary	Other Earnings	Total Earnings	Payout	Payout as % of Base	Payout as % of Total	Merit Rating	Merit Points
	William Dixon	$150,000	$45,000	$195,000	$2,258	1.51%	1.16%	B	10
	Lynda Harris	$100,000	$22,000	$122,000	$3,388	3.39%	2.78%	A	15
	Robert Campbell	$95,000	$15,000	$110,000	$1,129	1.19%	1.03%	C	5
	Neil Prior	$75,000	$0	$75,000	$1,129	1.51%	1.51%	C	5
	James Curtis	$72,000	$0	$72,000	$1,129	1.57%	1.57%	C	5
	Jeannette Totten	$70,000	$0	$70,000	$452	0.65%	0.65%	D	2
	Jack Kelly	$68,000	$0	$68,000	$2,258	3.32%	3.32%	B	10
	Ann Walker	$65,000	$0	$65,000	$1,129	1.74%	1.74%	C	5
	Nancy Dawson	$48,000	$0	$48,000	$3,388	7.06%	7.06%	A	15
	Thomas Whicher	$32,000	$0	$32,000	$2,258	7.06%	7.06%	B	10
	Lorrie Pierce	$32,000	$0	$32,000	$3,388	10.59%	10.59%	A	15
	Susanna Jackson	$25,000	$15,500	$40,500	$2,258	9.03%	5.58%	B	10
Hourly	111	$2,664,000	$222,000	$2,886,000	$125,345	4.71%	4.34%	C	555
Salaried	15	$330,000	$22,500	$352,500	$16,939	5.13%	4.81%	C	75
Sales Reps.	12	$240,000	$120,000	$360,000	$13,551	5.65%	3.76%	C	60
		$4,066,000	$462,000	$4,528,000	$180,000	4.43%	3.98%		797
				$4,528,000		4.43%			

		5.3

Payout Per Person for Multi-Incumbent Job Classes

Hourly =	$1,129
Salaried =	$1,129
Sales Reps. =	$1,129

Input

1. In Column I input the letter code corresponding to the performance level of the named individual. For the multi-incumbent job classes assume that the "average" performance level is C unless you have evidence to the contrary.

2. In Column C (Cells 5-9) input the description and letter code for the categories of performance that you wish to use. If you decide to use the categories shown in this example no input is necessary.

3. In Cells E5-E9 input the points you wish to allocate to the various performance levels.

4. In Column J you can either manually input the points corresponding to the Merit Rating in Column I OR place an @IF command (Quattro Pro 5.0) in that Column.

Tests

1. Divide the total of Column J by the number of employees (in this case, 150). With a normal distribution the average should be equal to, or slightly higher than, the number of points allocated to the "Meets job requirements" category. This calculation is shown in Cell J32 which shows, in this example, a result of 5.3.

Profit-Sharing Plan Worksheet

Option G1 Allocation = Equal Distribution

	A	B							Row Number
Fiscal Period	94-95				**Assumption**				4
Profits	$1,500,000				Number of Employees		150		5
% of Profits	12								6
Fund	$180,000								7
									8
	A	B	C	D	E	F	G	H	
Salary Grade/ Job Level		Name of Employee	Base Annual Salary	Other Earnings	Total Earnings	Payout	Payout as % of Base	Payout as % of Total	9–10
		William Dixon	$150,000	$45,000	$195,000	$1,200	0.80%	0.62%	11
		Lynda Harris	$100,000	$22,000	$122,000	$1,200	1.20%	0.98%	12
		Robert Campbell	$95,000	$15,000	$110,000	$1,200	1.26%	1.09%	13
		Neil Prior	$75,000	$0	$75,000	$1,200	1.60%	1.60%	14
		James Curtis	$72,000	$0	$72,000	$1,200	1.67%	1.67%	15
		Jeannette Totten	$70,000	$0	$70,000	$1,200	1.71%	1.71%	16
		Jack Kelly	$68,000	$0	$68,000	$1,200	1.76%	1.76%	17
		Ann Walker	$65,000	$0	$65,000	$1,200	1.85%	1.85%	18
		Nancy Dawson	$48,000	$0	$48,000	$1,200	2.50%	2.50%	19
		Thomas Whicher	$32,000	$0	$32,000	$1,200	3.75%	3.75%	20
		Lorrie Pierce	$32,000	$0	$32,000	$1,200	3.75%	3.75%	21
		Susanna Jackson	$25,000	$15,500	$40,500	$1,200	4.80%	2.96%	22
Hourly	111		$2,664,000	$222,000	$2,886,000	$133,200	5.00%	4.62%	23
Salaried	15		$330,000	$22,500	$352,500	$18,000	5.45%	5.11%	24
Sales Reps.	12		$240,000	$120,000	$360,000	$14,400	6.00%	4.00%	25
			$4,066,000	$462,000	$4,528,000	$180,000	4.43%	3.98%	26
	12								27
	150				$4,528,000				28
	138	**Input**			**Tests**				

Input

1. In Cell G4 input the total number of employees in the company.

Comments

1. Although the $ payouts are all equal and this appears equitable look at the wide range in Column G: Payout as % of Base; there is a multiple of over 6 times. Does this agree with the culture of your company?

Tests

1. The total number of employees you input into Cell G4 should be equal to the number in Cell C28. In Cell B27 the formula counts the number of individuals and in Cell B28 the formula adds the numbers of employees in each of the multi-incumbent job classes. Cell C28 is tht total of these two cells.

2. For each individual employee the payouts in Column F should be equal to each other and to the numbers in the Box-Payout Per Person For Multi-Incumbent Job Classes.

Payout Per Person for Multi-Incumbent Job Classes

Hourly =	$1,200
Salaried =	$1,200
Sales =	$1,200

Profit-Sharing Plan Worksheet

Option H1 Allocation = Length of Service/Base Earnings

	A	B	C	D	E	F	G	H	I
Fiscal Period	94-95								
Profits	$1,500,000								
% of Profits	12				**Assumptions**				
Fund	$180,000				Points per year of service =		2		
					Points per $1,000 of base =		1		
					Average years	Hourly	Salaried	Sales Reps	
					of service =	5	3	1	
Salary Grade/ Job Level	Name of Employee	Base Annual Salary	Other Earnings	Total Earnings	Payout	Payout as % of Base	Payout as % of Total	Years of Service
	William Dixon	$150,000	$45,000	$195,000	$5,361	3.57%	2.75%	5
	Lynda Harris	$100,000	$22,000	$122,000	$3,552	3.55%	2.91%	3
	Robert Campbell	$95,000	$15,000	$110,000	$3,451	3.63%	3.14%	4
	Neil Prior	$75,000	$0	$75,000	$2,714	3.62%	3.62%	3
	James Curtis	$72,000	$0	$72,000	$2,748	3.82%	3.82%	5
	Jeannette Totten	$70,000	$0	$70,000	$2,480	3.54%	3.54%	2
	Jack Kelly	$68,000	$0	$68,000	$2,614	3.84%	3.84%	5
	Ann Walker	$65,000	$0	$65,000	$2,245	3.45%	3.45%	1
	Nancy Dawson	$48,000	$0	$48,000	$1,876	3.91%	3.91%	4
	Thomas Whicher	$32,000	$0	$32,000	$1,273	3.98%	3.98%	3
	Lorrie Pierce	$32,000	$0	$32,000	$1,407	4.40%	4.40%	5
	Susanna Jackson	$25,000	$15,500	$40,500	$905	3.62%	2.23%	1
Hourly	111	$2,664,000	$222,000	$2,886,000	$126,456	4.75%	4.38%	555
Salaried	15	$330,000	$22,500	$352,500	$14,073	4.26%	3.99%	45
Sales Reps.	12	$240,000	$120,000	$360,000	$8,846	3.69%	2.46%	12
		$4,066,000	$462,000	$4,528,000	$180,000	4.43%	3.98%	653

Salary Points = 4066

Total Salary and Service Points = 5372

Total Service Points = 1306

Input

1. In Cell G4 input the number of points per year of service.
2. In Cell G5 input the number of points per $1000 of base salary.
3. In Row 7 (Cells F7, G7, H7) input the average number of years of service in each of the job classes.
4. In Column I input the number of years of service for each individual. Do NOT input anything in Column I for the multi-incumbent job classes (I23-25).

Comments

1. Try altering the balance between the effects of Earnings versus Service. For example, give 4 points per year of service (Cell G4) and see what happens. Try various other combinations until you feel the balance between the two factors is reasonably fair.

Payout Per Person for Multi-Incumbent Job Classes

Hourly =	$1,139
Salaried =	$938
Sales Reps. =	$737

Row Numbers: 4–32

89

Profit-Sharing Plan Worksheet

Fiscal Period	94-95
Profits	$1,500,000
% of Profits	12
Fund	$180,000

Option I1 Allocation = Length of Service/Job Level

Assumptions

Job Levels		%age of Fund	Number of Employees
Top Management = A		10	3
Middle Management = B		25	6
Supervisors/Professionals = C		15	3
All Others = D		50	138

	Within Category D		Average Years
Hourly	111		5
Salaried	15		3
Sales	12		1

Points/year of service: 6

Salary Grade/ Job Level	Name of Employee	Base Annual Salary	Other Earnings	Total Earnings	Payout	Payout as % of Base	Payout as % of Total	Years of Service	Service Points
A	William Dixon	$150,000	$45,000	$195,000	$7,500	5.00%	3.85%	5	30
A	Lynda Harris	$100,000	$22,000	$122,000	$4,500	4.50%	3.69%	3	18
A	Robert Campbell	$95,000	$15,000	$110,000	$6,000	6.32%	5.45%	4	24
B	Neil Prior	$75,000	$0	$75,000	$5,870	7.83%	7.83%	3	18
B	James Curtis	$72,000	$0	$72,000	$9,783	13.59%	13.59%	5	30
B	Jeannette Totten	$70,000	$0	$70,000	$9,783	13.98%	13.98%	5	30
B	Jack Kelly	$68,000	$0	$68,000	$9,783	14.39%	14.39%	5	30
B	Ann Walker	$65,000	$0	$65,000	$1,957	3.01%	3.01%	1	6
B	Nancy Dawson	$48,000	$0	$48,000	$7,826	16.30%	16.30%	4	24
C	Thomas Whicher	$32,000	$0	$32,000	$6,750	21.09%	21.09%	5	30
C	Lorrie Pierce	$32,000	$0	$32,000	$13,500	42.19%	42.19%	10	60
C	Susanna Jackson	$25,000	$15,500	$40,500	$6,750	27.00%	16.67%	5	30
D - Hourly	111	$2,664,000	$222,000	$2,886,000	$81,618	3.06%	2.83%	555	3330
D - Salaried	15	$330,000	$22,500	$352,500	$6,618	2.01%	1.88%	45	270
D - Sales Reps.	12	$240,000	$120,000	$360,000	$1,765	0.74%	0.49%	12	72
		$4,066,000	$462,000	$4,528,000	$180,000	4.43%	3.98%		

Input

1. In Column E, Rows 5-8 establish the number of Job Levels you wish to have along with their letter Code.
2. In Cells G5-8 input the percentage of fund you wish to allocate to each Job Level.
3. In Columns G and H input the types of Multi-incumbent Job Classes you wish and the number of employees in each. These should correspond to Columns A and B.
4. In Column I, Rows 10-12 input the average number of years of service for each Multi-incumbent Job Class.
5. In Column I input the number or years of service of each individual named in Column B. DO NOT input anything for the Multi-incumbent Job Classes. This is calculated automatically when you complete Step 4.
6. Input the number of points per year of service in Cell G13.

Tests

1. Try changing the number of points per year of service in Cell G13. The payouts per individual in Column F should NOT change.
2. For individuals in a Job Level with the same years of experience the payouts in Column F should be equal. See Curtis, Totten, and Kelly.

$4,528,000

Payout Per Person for Multi-Incumbent Job Classes

Hourly =	$735
Salaried =	$441
Sales Reps. =	$147

Profit-Sharing Plan Worksheet

Option J1 Allocation = Merit Ratings/Base Earnings

Fiscal Period	94-95
Profits	$1,500,000
% of Profits	12
Fund	$180,000

Assumptions

Merit rating	Points	Code
Outstanding =	15	A
Good =	10	B
Satisfactory =	5	C
Needs improvement =	2	D
Unsatisfactory =	0	E

Points per $1000 of base earnings = 2

A	B	C	D	E	F	G	H	I	J	Row Number
Salary Grade/ Job Level	Name of Employee	Base Annual Salary	Other Earnings	Total Earnings	Payout	Payout as % of Base	Payout as % of Total	Merit Rating	Merit Points	
	William Dixon	$150,000	$45,000	$195,000	$6,197	4.13%	3.18%	B	10	15
	Lynda Harris	$100,000	$22,000	$122,000	$4,298	4.30%	3.52%	A	15	16
	Robert Campbell	$95,000	$15,000	$110,000	$3,898	4.10%	3.54%	C	5	17
	Neil Prior	$75,000	$0	$75,000	$3,099	4.13%	4.13%	C	5	18
	James Curtis	$72,000	$0	$72,000	$2,979	4.14%	4.14%	C	5	19
	Jeannette Totten	$70,000	$0	$70,000	$2,839	4.06%	4.06%	D	2	20
	Jack Kelly	$68,000	$0	$68,000	$2,919	4.29%	4.29%	B	10	21
	Ann Walker	$65,000	$0	$65,000	$2,699	4.15%	4.15%	C	5	22
	Nancy Dawson	$48,000	$0	$48,000	$2,219	4.62%	4.62%	A	15	23
	Thomas Whicher	$32,000	$0	$32,000	$1,479	4.62%	4.62%	B	10	24
	Lorrie Pierce	$32,000	$0	$32,000	$1,579	4.94%	4.94%	A	15	25
	Susanna Jackson	$25,000	$15,500	$40,500	$1,199	4.80%	2.96%	B	10	26
Hourly	111	$2,664,000	$222,000	$2,886,000	$117,608	4.41%	4.08%	C	555	27
Salaried	15	$330,000	$22,500	$352,500	$16,193	4.91%	4.59%	B	150	28
Sales Reps.	12	$240,000	$120,000	$360,000	$10,795	4.50%	3.00%	C	60	29
		$4,066,000	$462,000	$4,528,000	$180,000	4.43%	3.98%		872	30

Total Earnings Points = 8132 $4,528,000

Earnings Plus Merit Points = 9004

Payout Per Person for Multi-Incumbent Job Classes

Hourly =	$1,060
Salaried =	$1,080
Sales Reps. =	$900

Input

1. In Cells F5-9 input the number of points you wish for each level of performance from A to E.
2. In Cell F10 input the number of points for each $1000 of base salary.
3. In Column I input the merit rating for each individual named. For the multi-incumbent job classes it is reasonable to assume that the average rating will be C or Satisfactory although in this example I have input B for the salaried group.

9
DEFERRED PROFIT-SHARING PLANS

This chapter discusses issues that are exclusive to designing deferred profit-sharing plans (DPSPs). These issues either do not occur, or are different for, cash and employee profit-sharing plans (EPSPs). They are

- membership
- vesting
- forfeitures
- withdrawals
- investment policy
- investment choice for the employee
- administration of invested funds
- employer contributions
- excess contributions
- methods of withdrawing funds from the plan when leaving the organization
- setting up a plan and registering it with Revenue Canada

MEMBERSHIP

The *Income Tax Act* specifies that the following persons cannot be members of a DPSP:[1]

- a person related to the employer;
- a specified shareholder of the employer;
- a person related to a specified shareholder of the employer; and
- a person related to a member of a partnership, where the partnership is the employer.

Generally, aside from these restrictions, you can follow the design guidelines in Chapter 6, "Membership in the Plan."

VESTING

My definition of vesting is "the irrevocable right of an employee to a portion of the employer's contribution to a profit-sharing trust (including earnings, capital gains/losses and forfeitures)." This means that if a dollar is placed in the trust fund of a profit-sharing plan, it belongs to the employee if it is "vested." If the employee leaves the plan (under any circumstances), he or she is entitled to that dollar. The employer does not have any right to it and is only holding it in trust for the employee. If the dollar is not vested, the employee does not have any right to that dollar if he or she leaves the plan.

Since 1991 the *Income Tax Act* has required that all employer contributions to a DPSP be vested with any employee who has been a member of the plan for two years or more.[2] This vesting requirement is quite different from those in effect before the revisions to the *Income Tax Act* in 1991 and readers are cautioned to ensure that they are using current reference material regarding vesting.

Although the revised 1991 rules are now more restrictive, the employer still has some choice about the vesting provisions of the new plan. The choices are immediate vesting, two-year vesting or somewhere in-between (this third option is not a realistic choice because of the narrow range of options available).

[1] *Income Tax Act*, Section 147 (2) (k.2).
[2] *Income Tax Act*, Section 147 (i) (ii).

Immediate vesting means that as soon as an employee becomes a member of the profit-sharing plan, any monies placed in the trust fund belong to him or her as soon as they are deposited. Thus, if the employee leaves the company after only three months, all contributions made to the plan go with him or her. If you choose the two-year vesting option, you start making contributions for each employee. But if an employee leaves after only one year, he or she cannot take those contributions made in his or her name; they are not yet "vested in him or her."

The rationale for using a longer period for full vesting is that it may encourage employees to stay with the company. If your initial objective for introducing a profit-sharing plan included reducing turnover or rewarding long service, it is more appropriate to use the two-year vesting rule. On the other hand, immediate vesting is preferable if you want employees to feel part of the company team and contribute fully to the company's operations. It eliminates the notion of being "second-class" members of the plan. Remember, if the plan has a long initial waiting period, such as a year, before the employee becomes a plan member, and uses the two-year vesting rule, it will be three years before the employee is a "full-fledged" member of the profit-sharing plan. Immediate vesting also has the advantage of needing less administration.

FORFEITURES

Forfeitures are another feature of DPSPs. Forfeitures occur when an employee leaves the plan and his or her funds are not fully vested. In such a case, the monies remain in the plan or trust fund. Obviously, this will not be a problem if you decide on immediate vesting.

According to the revised 1991 *Income Tax Act*, forfeitures must be taken back by the company.[3] Although they can be returned to the fund, they must be considered part of the company contribution for that year that is subject to the maximums imposed by the *Income Tax Act*. Following is an example based on already mentioned and new assumptions about our sample company in Chapter 7:

[3] *Income Tax Act*, Section 147 (2) (i.1).

- Lorrie Pierce is an engineer with 11 years of service and a base salary of $32,000 per year;
- the total cash payroll is $4,528,000 per year;
- Lorrie has $4,000 in unvested funds in the plan and leaves the company on the last day of the fiscal year;
- for this fiscal year the employer contribution formula will generate the maximum allowable amount (i.e., 18 per cent of $4,066,000, which equals $731,880).

In this case, the employer can leave the $4,000 in the fund and reduce the actual contribution to $727,880 ($731,880−$4,000) and still pay the amount determined by the contribution formula, which is $731,888. If, however, the contribution formula produced a result of $600,000, the employer would have two choices: add the forfeiture to that amount and actually contribute $604,000 to the plan, or reduce the calculated amount by $4,000 and contribute $596,000. Although this second option sounds somewhat devious, it still complies with the terms of the plan.

WITHDRAWALS

Another issue that needs to be addressed in designing a DPSP concerns withdrawals. Should employees be allowed to take money out of the deferred profit-sharing trust and, if so, under what circumstances? This is not an issue for employees who leave the company; they must withdraw the funds (see the section on page 102 called "Leaving the Plan"). This is an issue for employees who want to take monies out of the trust while still employed by the company.

Some people would argue that a deferred plan, by definition, should not allow withdrawals since its main purpose is to build tax-sheltered savings over the long term. Allowing withdrawals defeats the basic purpose of such a plan.

The opposite perspective is that deferral does not necessarily mean forever and there are times in an individual's life when withdrawals are appropriate. Such occasions include the purchase of a house; financing a child's education or in the case of a major illness.[4] Prohibiting withdrawals entirely can have the effect of forcing an

[4] As noted previously, DPSPs can be used as a form of long-term disability insurance.

employee to quit and leave the company in order to remove the money from the plan. This is obviously undesirable and a powerful argument for allowing withdrawals under certain circumstances.

Consequently, it is appropriate to review your original objectives in establishing the profit-sharing plan (see Chapter 5) when deciding how to handle withdrawals. If your objective was to provide retirement income (either instead of, or in addition to, a pension plan), withdrawals will be prohibited. On the other hand, if your objective was to provide additional benefits, such as disability insurance, you must allow withdrawals.

Many companies compromise by allowing withdrawals, at least under certain circumstances, but often with restrictions and/or penalties. The issues facing the committee at this stage of the design process are to specify:

- the permissible reasons for withdrawals;
- any restrictions;
- penalties; and
- who makes the decision to allow a specific withdrawal.

With the exception of the final item, these are highly subjective decisions. Consequently, I recommend that you, as the employer, get a lot of input from your employees since it is important that the rules established reflect their values.

The usual reasons companies allow withdrawals have been listed above. It is also possible to decide that employees do not need a reason to make a withdrawal and that the company can exercise control through the use of penalties and restrictions. This is administratively easier but if you are considering making withdrawals this easy, perhaps you should consider whether you need a DPSP at all. Restrictions can take a wide variety of forms but some examples are as follows:

- limit the number of withdrawals per lifetime;
- place a maximum on the percentage of vested funds that can be withdrawn;
- restrict the withdrawal to monies that have been vested for a minimum period;
- establish a maximum dollar amount; and

- restrict withdrawals to fully vested funds — this would not apply if you have immediate vesting.

These restrictions can be used individually or in combination. Similarly, penalties can vary widely. Two examples are as follows:
- suspension of company contributions for a specified period such as one year; and
- prohibiting of future withdrawals for a period of time.

You should carefully consider before implementing any restrictions whether it is fair to penalize employees who make withdrawals, presumably for sound reasons. On the other hand, some employers use penalties to encourage the concept of deferral. However, it may be more appropriate to control withdrawals by specifying the allowable reasons for, and restrictions on, them. Remember that the *Income Tax Act* imposes its own penalties in that all withdrawals are taxed at full marginal rates.

If you do decide to allow withdrawals, but only for specified reasons, you must develop a process to review any request and ensure that it meets these criteria. In most cases this will be a very straightforward process. However, areas of ambiguity can arise. For example, suppose that your plan allows withdrawals for educational purposes and the employee intends to use the withdrawn funds to send his or her child on a world tour. Does this reason satisfy this criterion? The profit-sharing committee usually is charged with making this decision. However, employees may be reluctant to make these types of decisions about their colleagues and it may be more appropriate if the owner/manager or some other senior executive took on this responsibility.

INVESTMENT POLICY

Company contributions to a DPSP are placed in a trust fund and consequently the company needs an investment policy regarding how the funds should be invested. Investment policy has three aspects:
- the restrictions imposed by the *Income Tax Act;*
- where you will actually invest the funds placed in the trust; and
- the question of employee choice of investments.

INCOME TAX ACT

Section 204(a-i) of the *Income Tax Act* defines what "qualified investments" for a DPSP are. In general, these include all forms of investment such as cash, GICs, equities, bonds, coupons and mutual funds, with two major exceptions. DPSPs cannot invest in debt instruments (e.g., bonds) of the employer company or shares of a privately owned corporation.

However, the plan can invest in shares of a publicly owned corporation including the employer company.

WHERE TO INVEST

Although the *Income Tax Act* is somewhat limiting in terms of where the plan can invest, there are still a number of major options available.

For example, you can decide whether you are a conservative or an aggressive investor by answering the question: are you concerned about income or growth? Remember that since the DPSP is a tax-sheltered plan (like an RRSP), many of the same investment guidelines apply.

A second issue is that you must decide whether to invest, either wholly or partially, in your own company shares. As mentioned above, this is only possible in a public company. This is a "double-edged sword" with major advantages and disadvantages. The advantage is that the financial results of the efforts of the employees are compounded: the better the company does, the more money goes into the plan and the company shares in the plan become more valuable.

However, this is also the major flaw, especially if the profit-sharing plan is the only retirement vehicle for the employees. Since many companies provide deferred profit-sharing plans instead of registered pension plans, providing retirement income is often the plan's major objective. However, this creates a situation of "putting all your eggs in one basket." If the stock price happens to be at a low point when an employee retires (and this will happen to all companies sooner or later), pension payments can be seriously affected.

Employee Choice of Investments

Many plans are designed so that employees can choose where their monies will be invested while still within the DPSP. For example, a plan may have three sub-plans with risk factors of low, medium and high. Plan A could be all government bonds; Plan B could be a balanced mutual fund; and Plan C could be high-risk but also high-potential equities. Such an arrangement allows younger employees to take a chance on Plan C while older employees within a few years of retirement could choose Plan A and be reasonably certain of what funds will be available when retirement actually occurs.

The advantage of giving employees a choice of investments is similar to that of combined plans (i.e., cash and deferred) in that the plan can then meet a wider variety of employee needs. The disadvantage is that it involves more complications and costs, and more investment decisions need to be made.

Since few employees are sophisticated investors, the chief executive who is interested in allowing employees various choices of investments must address the question of what obligation the company is willing to undertake for the quality of the choices that employees make. Will employees be allowed complete discretion? Will they be provided with professional advice and, if so, by whom? Or will they only be allowed very limited options?

Besides the actual sub-plan choices that employees may be offered, the design committee should also address the issue of when such choices can be made. There are two major options: at the point of allocation and transfer of current balances at specified times.

Employees can be allowed to choose which sub-plan will be used each time that funds are allocated to the plan. This means that each time there is a distribution of profits, the company would have to determine each employee's decision. This can be quite an onerous task for the plan administrator if the company employs a large staff or if distributions are made more often than annually. A common way of handling this is to place all funds, on initial allocation, automatically in one of the sub-funds, usually the most conservative one. This is a "default" position that will be taken unless the individual employee takes the initiative and advises the plan administrator of a different selection.

The other option is to allow employees to transfer their account balances at various times such as every five years, at certain ages, or to limit the number of transfers during the employee's service with the company. An important time that should be permitted is at five and 10 years before retirement. This is a critical period because the employee near retirement is, in effect, running out of time and needs to be sure that his or her investments are congruent with retirement plans.

ADMINISTRATION OF INVESTMENTS

Any profit-sharing plan (deferred or cash) should have a committee that is responsible for ongoing administration of the plan (see Chapter 12). If the plan is a DPSP or an EPSP, this committee should include investments as one of its major responsibilities. This does not mean, however, that committee members have to become investment specialists; these specialists can be hired. However, the committee should probably include the chief financial officer or treasurer of the company.

There is another issue of membership in the committee — that is, whether rank-and-file employees should be included in the investment process. Most employees — including company executives — are not experienced investors. Furthermore, many employees do not want the responsibility of making investment decisions about money that belongs to their colleagues, even with the help of professional advisers. One way to deal with this situation is to have a separate investment committee comprised of the company president, the chief financial officer and the chair of the main committee.

Legislation requires that trustees of a DPSP are responsible for conducting their investment decisions and management of the trust according to the "prudent person" rule. In general, this means that they conduct themselves the way a "prudent person" would but more specifically requires

1. that all securities be bought and sold at fair market value;
2. that the rate of return is competitive with other, similar investments; and
3. that they maintain sufficient liquidity in order to meet the foreseeable obligations of the funds such as retirements, resignations and withdrawals.

In the early years of a plan, when the fund is relatively small, it is probably easier to invest in extremely safe vehicles such as government bonds or GICs, which can be easily handled by an in-house group. However, as the fund grows, it is better to obtain professional advice that is available from a wide variety of sources including investment brokers, banks and trust companies. Since terms can vary widely, a prudent committee will "shop around" for the best deal.

LEAVING THE PLAN

The final decision regarding deferred profit-sharing plans is not really a design issue; it is merely that plan designers should be aware of the options available to each member of a DPSP when the member retires or leaves the plan for any other reason such as death, disability, and voluntary or involuntary termination. There are five major choices:

- lump-sum single payment[5]
- in equal instalments payable not less frequently than annually over a period not exceeding 10 years;[6]
- purchase of an annuity;[7]
- transfer to an RRSP, a RPP or another DPSP;[8] and
- take the payment in company stock.[9]

If the first option is chosen, the payment is treated as cash and taxed at full personal rates. Obviously, with a large sum of money, this could be quite onerous but there are situations in which it may meet the wishes of the employee who needs a large amount of cash for reasons such as the purchase of a retirement home or a business.

The employee may also remove the funds in the plan in equal payments over a period not to exceed 10 years. The payments are taxed in the same way as the lump-sum option. If the employee is younger than age 71, the total sum can be transferred to an RRSP, another DPSP or a registered pension plan.

[5] *Income Tax Act*, Section 147 (10).
[6] *Income Tax Act*, Section 147 (2) (k) (v).
[7] *Income Tax Act*, Section 147 (2) (k) (vi).
[8] *Income Tax Act*, Section 147 (19).
[9] *Income Tax Act*, Section 147 (10.1) and 147 (10.2).

An annuity can also be purchased. If the funds are transferred directly, no tax is payable at the time.

The final option is only available if the plan holds stock in the employer company (note that since a DPSP cannot hold stock in a private company, this option is only available to public companies) and the employee receives actual stock rather than cash. In this case, capital gains tax is payable but only when the employee sells the stock. The capital gains tax is payable only on the amount gained between the time the shares were received from the plan by the employee and when they are sold.

CONTRIBUTIONS

The 1991 amendments to the *Income Tax Act* made significant changes to the amounts that employers were allowed to contribute to a DPSP and how these limits were calculated. Briefly, the amendments eliminated employee contributions to a DPSP. Previously, employees had been allowed to contribute (some employers made this a mandatory contribution) to a deferred plan and, although these contributions were not deductible, the earnings and capital gains that they accrued while in the plan were not taxable.

The other change is that DPSP limits used to be separate from other retirement vehicles such as registered pension plans and RRSPs. They are now integrated and contributions to the three types of retirement vehicle are treated as a total.

The new rules are outlined in section 147(5.1 and 5.11) of the *Income Tax Act* and in associated regulations and interpretation bulletins.

The total contributions by both the employee and employer are called the "money purchase limit." The amount that an employer can contribute to a deferred profit-sharing plan must be the lesser of one-half of the money purchase limit or 18 per cent of the employee's compensation. In 1995, the amount permitted is $6,250. For 1996 and subsequent years, the amount will be indexed.

EXCESS CONTRIBUTIONS

Since the employer is limited to the amount of funds that can be placed in a deferred plan, the design committee must decide what to do if the employer's formula for determining the contribution to the plan results in an amount that exceeds the limits in the *Income Tax Act*. This can only happen if the employer contribution is defined (e.g., five per cent of net profits before tax) unlike a discretionary formula, where management makes a different decision every year about the amount to be paid into the profit-sharing plan.

If the employer contribution exceeds the allowable limits, there are several options depending on whether the DPSP is the primary or secondary part of the combination plan (see Chapter 2 for a discussion of the advantages and disadvantages of combination plans).

When the DPSP is the primary part of the plan, the plan was probably set up to provide retirement income. Therefore all funds, up to the allowable limits, are contributed to the deferred plan. Suppose the allocation formula resulted in an $8,000 payment to the employee. Of this amount, $6,250 would be placed in the DPSP, while the excess of $1,750 can be paid in cash or put into an EPSP. A cash payment is obviously the easier option since using an EPSP means that the employer must set up such a plan, which increases the administrative complexity of the plan.

The opposite situation can also occur. Suppose that the combination plan specifies that, for example, 50 per cent of the funds generated by the formula are first paid in cash and the remainder goes to the DPSP but this amount still exceeds the allowable limits. In this case, the excess should just be paid in cash.

SETTING UP A DEFERRED PROFIT-SHARING PLAN

Once you have made all the design decisions discussed above, it is necessary to set up the trust and register the plan with Revenue Canada.

The trust fund can be established either with a Canadian trust company (provincially or federally incorporated) or the employer can establish a separate trust, which must have at least three individual trustees. Each trustee must be a resident of Canada and at least one of them cannot be a shareholder of the employer company and must be independent of the employer, i.e., someone whose relationship with the employer is at arm's length.

10
EMPLOYEE PROFIT-SHARING PLANS

The concept of an employee profit-sharing plan (EPSP) is addressed in section 144 of the *Income Tax Act*. As mentioned earlier, this term can create confusion since all profit-sharing plans discussed in this book involve employees.

Many of the issues that are involved in designing an EPSP are almost the same as those regarding a DPSP (see Chapter 9), although the treatment is often quite different. These issues are

- membership
- vesting
- forfeitures
- withdrawals
- investment policy
- investment choice for the employee
- administration of invested funds
- employer contributions
- employee contributions

- methods of withdrawing funds from the plan
- when leaving the organization
- setting up a plan

MEMBERSHIP

The *Income Tax Act* places no restrictions on who can be members of an EPSP; therefore you can follow the guidelines laid out in Chapter 6, "Membership in the Plan." Remember that if the EPSP is part of a combination plan, the two parts can have different membership rules. This combination is not typically recommended since it would add considerably to the complexity and difficulty of communication and administration. It is particularly inappropriate if the EPSP is for the purpose of holding "excess" contributions to a deferred profit-sharing plan. Refer to the "Excess Contributions" section in Chapter 9.

VESTING

The definition of vesting is still "the irrevocable right of an employee to a portion of the employer's contribution to a profit sharing trust" (including earnings, capital gains/losses and forfeitures) as in Chapter 9 for DPSPs. *The Income Tax Act* does not specify any vesting requirements for EPSPs. The choice is between immediate vesting and graduated vesting.

Immediate vesting means that as soon as an employee becomes a member of the profit-sharing plan, any monies placed in the trust fund belong to him or her when they are deposited. Immediate vesting is preferable to graduated vesting if you want employees to quickly feel part of the team and to contribute fully to the company's operations. Immediate vesting eliminates the notion of being "second-class" members of the plan.

However, if your initial objective included reducing turnover or rewarding long service, it is probably more appropriate to use graduated vesting. In Chapter 9, graduated vesting was dismissed as an option for DPSPs because the two-year rule in the *Income Tax Act* narrowed the options so dramatically. This is not a problem with EPSPs. Here is an example of a graduated vesting schedule.

GRADUATED VESTING SCHEDULE

PERCENTAGE OF FUNDS	YEARS OF SERVICE
10	1
20	2
40	5
100	10

This type of schedule can illustrate two types of graduated vesting: class year vesting and total fund vesting. If this were class year vesting, in any given year an employee with two years' service would have 20 per cent of the allocation to his or her account each year vested. If he or she had 10 years of service, 100 per cent of the allocation each year would be vested.

With total fund vesting, the same employee would be entitled to 20 per cent of all funds allocated to him or her in the plan if he or she left the company after two years of employment. If the employee left the company after 10 years, he or she would be given 100 per cent of all monies in the plan that had been allocated to him or her plus the gain in their value. These examples of graduated vesting clearly show that total fund vesting is easier to communicate and administer.

There are however, two other disadvantages to using either type of graduated vesting. First, they greatly complicate the process of communication. Vesting is a difficult concept and even more difficult to explain to the average employee. One of the main principles of designing any human resource program — including profit sharing — is to keep it short and simple. Second, they create two types of plan members — those whose funds are completely vested and those whose funds are less than completely vested.

However, despite these cautions about graduated vesting, do not disregard this option if one of your objectives is to reduce turnover and/or reward longer service. Graduated vesting can be a very useful tool to achieve these goals.

Vesting rules do not apply to contributions by employees or any increase in their value. If the employee leaves the company, he or she is entitled to a refund of all of his or her contributions to the plan plus any accumulated interest and/or capital gains.

FORFEITURES

Forfeitures are defined the same for EPSPs as they are for DPSPs: funds that are left in the plan when employees who are not fully vested leave the company. Obviously, if you choose to use immediate vesting, forfeitures are not an issue.

Although they are defined the same way, forfeitures for EPSPs are treated quite differently than for DPSPs. The *Income Tax Act* is quite specific about how to treat forfeitures in a DPSP but does not address the issue for EPSPs. Consequently, you can handle forfeitures in two ways: the company can take them back or they can be reallocated to the remaining plan members.

Reclamation of the forfeitures by the company obviously saves the employer money and does not impose a cost on the remaining members of the plan. However, forfeitures also present an opportunity to further reward employees who stay with the firm without any additional cost to the employer since the money has, in a sense, already been spent. This requires that the funds be left in the plan and reallocated to the remaining members.

Reallocation can be done in two ways: in proportion to account balances or according to the original allocation formula.[1] An example of the former will illustrate the point. Assume that the EPSP fund has a value of $500,000; our sample employee has had $33,000 allocated to her account over her period of employment; and the total forfeitures in 1993 amount to $12,000. Using these account balances her share will be

$$\frac{33,000}{500,000} \times 12,000 = 6.6\% \times 12,000 = \$792$$

This amount will be credited to this employee's account — her total credits will now be $33,792.

Reallocating forfeitures according to account balances has the advantage of further reinforcing length of service since employees with longer service will usually have higher account balances. If the allocation formula uses earnings or job level, this method also further rewards those with higher salaries or positions. There is nothing wrong with this approach if that is what you choose to do. However, you (or your profit-sharing committee) may believe that

---

[1] See Chapter 7, "Allocation: Dividing the Funds."

having earnings in the basic allocation formula is going far enough and therefore the plan should probably use the basic allocation formula rather than account balances.

WITHDRAWALS

The arguments for or against withdrawals are the same for EPSPs as they are for DPSPs. The difference lies in the fact that, with EPSPs, income tax has already been paid and therefore withdrawals are not as onerous for the individual.

INVESTMENTS

Investments are somewhat different for EPSPs than for DPSPs since the funds in the plan are treated as though they belong to the employee. This means that investments should be made according to the tax rules that would apply if the investments were made by the employee personally. In other words, the funds should be placed in dividend-paying stocks rather than interest-bearing instruments since dividends are subject to a lower rate of taxation than interest, assuming that the level of risk is acceptable. The investments should also emphasize growth because of the capital gains exemption.

Employee profit-sharing plans have their biggest advantage over DPSPs in terms of allowable investments. EPSPs are permitted virtually any form of investment including three that are specifically forbidden for DPSPs: foreign securities, shares in a Canadian controlled private corporation (CCPC) and debt instruments of the employer corporation. Investing in foreign securities is of particular interest because many Canadian companies are wholly owned subsidiaries of foreign multinationals whose shares are not traded on Canadian stock exchanges. If the objectives of the profit-sharing plan include the encouragement of ownership, this can be one way to achieve this goal. It is a somewhat indirect means of doing so, however, since the shares owned are those of another company, the parent corporation, but it may be the only way to achieve this objective.

Having the EPSP own debt (i.e., bonds or debentures) of the employer provides an interesting way for the corporation to raise debt financing if it is having trouble doing so in more traditional ways. In effect, the company could provide at least part of its own financing and have an employee profit-sharing plan as well. Of course, the employer should ensure that the interest rates paid are competitive in the general market. The disadvantage of this approach is that the interest income will be taxed at a higher rate than if the funds were invested in dividend-paying shares with growth — and therefore capital gains — potential.

Buying shares in a CCPC is another option for the EPSP. Obviously this would usually only apply to the employer company. There would be no particular advantage to buying shares in another CCPC and several disadvantages. The pros and cons of using a profit-sharing plan to buy company stock (either public or private) have already been discussed in Chapter 9. These arguments apply equally to EPSPs.

INVESTMENT CHOICES FOR THE EMPLOYEE

The main advantage of EPSPs over DPSPs is that EPSPs can invest in the shares of a private company. Many people perceive this as the only advantage. If you set up the EPSP to buy stock in your own company AND be the retirement vehicle for your employees, investment choice for them is only if there is no pension plan (i.e., the EPSP is the only retirement fund they have) and, they are near retirement. Since there is no pension plan they are entirely at the mercy of the price of your company stock for their standard of living during retirement.

Therefore, the plan probably should provide an alternative investment such as guaranteed investment certificates or government bonds, to which the employee can switch either five or 10 years before retirement age. This way his or her retirement income will not depend entirely on your company's stock.

The other issues discussed in Chapter 9 about employee choice of investments are the same here: number and type of investment choices; timing of choices; and investment training of employees.

EMPLOYER CONTRIBUTIONS

Excess contributions are not an issue for EPSPs since the *Income Tax Act* does not impose any limits in terms of the maximum contribution to the fund.

EMPLOYEE CONTRIBUTIONS

Some EPSPs require employees to make contributions to the plan, similar to mandatory contributions to a pension plan. However, these contributions are not tax-deductible as they are when made to a registered pension plan.

If the purpose of the profit-sharing plan included the encouragement of employee savings, then mandatory employee contributions which can accomplish this goal should be considered. It can be further encouraged with the allocation formula by making at least part of the allocation related to the size of the employee contribution. For example, if the employee puts one dollar into the fund, the company could match that dollar. Alternatively, if you are using a combination formula, each $100 contribution could be given one point and each year of service (perhaps with a maximum) could receive two points. The overall fund would then be distributed in proportion to the number of points that each employee has. Compulsory employee contributions should only be used if there is a pension plan.

ON LEAVING THE PLAN

When an employee leaves an employee profit-sharing plan, he or she can receive the funds in a lump sum but, since taxes have already been paid, the employee will receive the total allocation. This is essentially the only option available since the others that apply under DPSP rules, such as transferring to an RRSP, are not applicable.

SETTING UP AN EMPLOYEE PROFIT-SHARING PLAN

EPSPs do not have to be registered with Revenue Canada. However, you must set up a trust fund, either with a Canadian trust company or in a separate trust established by the employer.

11 COMMUNICATIONS

This chapter discusses issues related to communications about the profit-sharing plan including timing, content and methodology.

Communication is the most vital part of any profit-sharing plan and can occur at many stages. We will divide communications into two major groups: those that happen during the design process and those that occur after the plan has been implemented.

COMMUNICATIONS DURING DESIGN OF THE PLAN

The main communications component during the design of the plan is involving the employees in the process described in Chapter 5. Employee participation serves as much a communications instrument as any other purpose, perhaps even more so, and it has the advantage that committee members provide a two-way communications channel. This two-way communication is vital if the company is to have a truly effective plan that employees are actively involved in developing, and that will not be viewed as imposed by the owners or executives.

During the design period, communications take place in three parts. The first occurs at the start of the project. The chief executive officer, or another appropriate official, should announce, in writing to each employee, the decision to develop a plan, the reason(s) for this decision and the objective(s) of the plan. Please see Chapter 5, page 39 for a sample announcement memo. If you have a company newsletter, this announcement should also be included in the next issue. It is also useful to post the announcement on any company bulletin boards.

During the design phase, periodic progress reports should be provided to all employees at significant points. These reports could include the selection/appointment of the committee; the results of the first meeting of the committee; a description of how the committee will be contacting employees and what questions will be asked; and, finally, when the plan has been approved and is about to be implemented.

COMMUNICATIONS AFTER IMPLEMENTATION

Once the plan has been implemented, it is important to ensure that communications about its progress become a routine part of company operations and the activities mentioned above. After implementation, communications about the profit-sharing plan can occur at the following times:

- during the recruitment process
- during the orientation of new employees
- when employees become members of the plan
- during supervisory training
- during meeting(s) of members of the plan
- when cheques are handed out
- when regular pay cheques are distributed
- as part of retirement planning

Recruiting

The best time to start the communications process is at the beginning of the relationship between the company and the employee.

The plan can be used as a selling tool when recruiting new employees: refer to it in your advertising, tell employment agencies about it and, most importantly, ensure that interviewers mention it to prospective candidates — especially those whom you are particularly interested in. Remember, with a broad-based profit-sharing plan, you are an elite employer in Canada and have every right to be proud of and promote this fact. When you make a written offer or confirmation of employment, list the profit-sharing plan as one of the features of employment.

Orientation

Regardless of whether your company has a formal orientation program, each new employee must be listed on the company's payroll. To do this, the employee must complete various forms regarding benefits such as group life insurance, pension plans and income tax. During this process, the profit-sharing plan can also be reviewed briefly and the booklet describing the plan can be given to the employee. This should be done even if the employee will not be a member of the plan for several years (see Chapter 6, "Membership in the Plan,"). If one of your objectives in establishing the plan was to reduce turnover, keep in mind that employees will not be motivated to remain with the company long enough to become members if they don't know about the plan or don't understand it.

On Joining the Plan

Almost all plans require new employees to complete a specified period of employment before becoming a member of the plan. This period is typically between three months and three years. The date on which the employee becomes a member of the plan provides another excellent opportunity for communications. Obviously, if there is no waiting period to join the profit-sharing plan the orientation interview discussed above should include a detailed review of the plan.

If more than one employee becomes eligible to join the plan in a given period (for example, during April), you can arrange for a group orientation session. However, try to have the indoctrination meeting as close to the actual date of joining the plan as is practical. The employee(s) should meet with the plan administrator (a half-hour meeting will usually be sufficient), who will explain the plan in detail and answer any questions. If the company has an employee newsletter, the names of the new members of the profit-sharing plan should be printed in the next issue. Alternatively, employees' names can be posted on bulletin boards, especially if certain bulletin boards have been designated for news about the plan.

Supervisory Training

First-line supervisors play a critical role in communicating the profit-sharing plan to employees. They are usually the people whom employees will turn to first if they have questions about the plan. Consequently, supervisors should be thoroughly trained in all aspects of the plan and should also be given extensive instruction in how to deal with suggestions, comments and observations about ways to improve profitability. Chapter 15 addresses this aspect of communications and suggestion plans in more detail.

Employee Meetings

Many employers hold an annual meeting for all plan members; if the company has two or more locations that are geographically dispersed, several meetings may be necessary. It is also possible to have quarterly, or even monthly, meetings, after the financial results for the period are available. In general, more frequent meetings provide the advantage that employees will have an improved understanding of the company's financial position and, if it is unsatisfactory, they can possibly take corrective action. The disadvantage of annual meetings is that the employees only see the results of their efforts after it is too late to do anything.

Cheques

This section reviews the distribution of cheques from two perspectives and the opportunities for communications connected with each perspective.

The first, and most obvious, is the distribution of the actual profit-sharing cheque. The frequency of this distribution will obviously depend on how often you calculate the profit-sharing fund.[1] This is a great opportunity to communicate about the plan or even to sell it. Even if your regular payroll is directly deposited to employees' bank accounts, I recommend that you do not direct-deposit the profit-sharing cheques. Instead, deliver the cheques directly to employees. If the company is small enough, the owner/manager can do this personally. Otherwise, the department manager or supervisor can deliver them. One final point: deliver the cheque itself, not just the cheque stub.

The second opportunity for communication involves regular pay cheques. Regardless of whether you use direct bank deposit, you must still provide employees with a statement of earnings that shows gross earnings including base pay, overtime, commissions, etc., and all deductions such as income tax, CPP and the employee's share of other benefit costs.

You can include information about the profit-sharing plan with the statement of earnings. Such information could range from the traditional "payroll stuffers" to corporate financial statements, or even, in the case of deferred plans, statements of investment performance.

Using the pay envelopes has two main advantages. First, you know that the employee receives the information. Whether he or she actually reads it is largely beyond your control. Second, since payrolls are done relatively frequently, you have the opportunity to generate information often and regularly for the members of the profit-sharing plan.

PRE-RETIREMENT PLANNING

The final time when communications about the plan are important is during the last few years before retirement (but only for DPSPs and EPSPs). If you have a formal retirement-planning process, discussions about the profit-sharing plans, and the payouts, can be built into this process. Employees will have to be advised on the options available to them for removing their vested funds from the plan.

---

[1] See the section called "Frequency and Timing of Payments" in Chapter 5.

If you have allowed employees to choose how their funds will be invested, this process of retirement planning can be started as early as 10 or 15 years before the normal retirement date so that, if the employee does switch investments, the funds will have time to grow.

The most obvious time to communicate with employees is when financial results are available. Most companies receive monthly financial statements and consequently should be able to advise staff at least 12 times a year. Do not leave communication of financial results until the end of the fiscal year. If monthly reporting is too onerous, provide quarterly reports instead. Remember, the power of the profit-sharing phenomenon depends on employee involvement; if they are unable to track their progress (and the company's progress) fairly regularly, they will not be able to respond until it is too late.

WHAT TO COMMUNICATE

Another issue to consider concerns which financial information to provide. There are widely divergent views on this subject, especially in privately owned companies that have a history of little or no disclosure of financial results. In many cases, the owners (i.e., family members) believe that the financial results are confidential and they would prefer not to, or flatly refuse to, disclose this information. On the other hand, some companies are willing to disclose almost everything in line with the philosophy that since the employees share the efforts and the results, it is appropriate to share information with them as well. This is a subjective decision; there is no "scientific" way to settle this issue.

If you are reluctant to share detailed financial information, you can still provide some data to employees that will help them to understand the company's economic performance and their effect on it.

First, you can make use of ratios. All numbers on a normal income statement can be expressed in terms of a percentage of gross revenue. For example, consider the following document:

INCOME STATEMENT – XYZ COMPANY
(for the year ending August 31, 1994)

Gross revenue	100.0
Less returns	5.0
Net sales	95.0
Cost of goods sold	60.0
Expenses	30.0
Gross profit	5.0
Income tax	2.5
Net profit	2.5

Naturally, the actual document would have much more detail but even a simplified statement like this can show the staff how they can affect profitability. The sales returns figure tells them that goods are being returned, usually because they are not satisfactory. This example shows that the cost of raw materials is significant and that, by using these materials more carefully, they can have a major effect on the bottom line and on their own profit-sharing award.

Another variation on the above is to show year-over-year comparisons using a similar method, as shown in the following example:

INCOME STATEMENT – XYZ COMPANY
(for the years ending August 31, 1994 and 1995)

	1994	1995 results as a % of 1994
Gross revenue	100.0	106.0
Less returns	5.0	101.2
Net Sales	95.0	105.6
Cost of goods sold	60.0	110.7
Expenses	30.0	98.4
Gross profit	5.0	99.1
Income tax	2.5	99.1
Net profit	2.5	101.9

In this example, the 1995 results are expressed as a percentage of the 1994 results in the same category. Gross revenue in 1995 is 106 per cent of what it was in 1994. Communicating results this way will enable employees to see and evaluate year-to-year changes.

Although these methods can be used together, you should be careful to explain that the ratios, although apparently similar, are in fact quite different.

Another option is to show financial results to a selected group of employees that you trust will not disclose the actual numbers to the other staff. If these employees are respected and trusted by most employees, their assurance that the financial results and the profit-sharing payments are consistent with your past practices and stated objectives for the plan will often be sufficient to reassure them. This group of trusted employees can be selected for this purpose or they could be the permanent profit-sharing committee. This group is usually not expected to communicate any information other than the fact that they are satisfied that the profit-sharing results are consistent with the actual financial results of the company.

A variation on this approach is to have the company's external auditors verify, in a written opinion sent directly to the employees, that all activities have been conducted properly.

COMMUNICATIONS MEDIA

The final point about communications is the need for a document describing the major features of the plan. Although this can be a professionally produced publication with glossy paper and expensive artwork, this is not necessary. It can be a two-or three-page document that is photocopied on company letterhead. The important point is that it should be a complete report that is written at a level of language suitable for the intended audience. In many companies in metropolitan areas with large immigrant populations, it may also be appropriate to translate it into other languages. Suggested chapter topics are:

- introduction (including a brief history of the plan)
- objective(s) of the plan

- company contributions
- membership criteria
- allocation formula
- investments (for DPSPs and EPSPs)
- administration of the plan (committee, etc.)
- investment choices available to employees (for DPSPs and EPSPs)

This document should be short and simple; one company describes the whole plan on one page. It is also essential that it be kept up-to-date. Very few plans remain the same over their entire life; consequently, as changes are made, it is critical that your communications program be updated.

This document should be given to employees either when they join the company or when they join the profit-sharing plan, or possibly at both times if the waiting period is several years.

If your plan includes a large enough number of employees to justify the cost, you can also investigate the use of videos or even CD-ROMs to communicate the plan or the periodic results. These can be extremely powerful communications vehicles. Employees could view these at the office, factory or warehouse or even be encouraged to take them home to share with their family.

The special communication requirements of deferred profit-sharing plans and employee profit-sharing plans are discussed in Chapters 9 and 10 respectively.

12
ADMINISTRATION

This chapter covers the issues regarding the ongoing administration of the profit-sharing plan once it is installed. Although the procedures discussed are not put in place until after the plan is implemented, they are part of the decision-making process of design and should be included in the deliberations of the profit-sharing committee. These issues are

- whether there should be a permanent profit-sharing committee;
- the composition and size of the committee (members and chair);
- term(s) of office of the committee members;
- selection of the committee;
- the transition from the present committee to the new committee;
- exact powers of the committee;
- responsibility for administrative functions;

- formal reviews of the plan, including opinion surveys of the employees; and
- administration of trust accounts.

SHOULD THERE BE A COMMITTEE?

The first issue concerns whether there should be a permanent[1] profit-sharing committee. The arguments for an ongoing committee are essentially the same as the reasons for a design committee that were presented in Chapter 5, although these arguments are probably more powerful at this point. If you decided to have a committee for the design process, it should not be difficult to decide to have an ongoing committee. Hopefully, the experience has convinced you of the usefulness of such a body. An ongoing committee will ensure that employees have continuing involvement and participation in the administration of the profit-sharing plan. This is important for two reasons.

First, profit-sharing plans are not static. They must evolve as the organization changes. The need for, and timing of, this change is unpredictable. This means that there is a continuous need for input from plan members, and a permanent committee is the most effective way to obtain this input.

Second, employees must be involved in the profit-sharing plan for it to be effective. Although this argument is similar to the previous argument, it is not the same. The first argument is that a committee provides a direct benefit to the employer/owner/manager. This second reason for a committee is that it provides a primary benefit to the employees, and a secondary benefit to the employer. By being involved in the plan through a committee, employees will have greater faith in its fairness and will be more committed to it.

COMPOSITION AND SIZE OF THE COMMITTEE

In Chapter 5, I recommended that the design committee comprise five to 10 members, although sometimes they may need to be larger in order to fulfil all the criteria listed. The permanent committee

---

[1] "Permanent" refers to the committee itself, not necessarily the membership therof.

does not need to be this large. Practical considerations such as the number of employees that can be released from their regular jobs for committee work and the ideal size of a workable committee indicate that a lower number is preferable. I believe that five committee members, plus the chair, is the most appropriate size for the permanent body. A smaller group is acceptable for the transition period since it is not as important at this point to disseminate a large amount of information and to obtain employee input and response. In effect, the workload is smaller and therefore fewer people are needed.

In determining the size of the committee, although it is still important to try to meet the criteria set out in Chapter 5, you can achieve this in broader terms. For example, instead of having a representative for each of the production, warehouse, production engineering and quality control departments, you could have one representative for the whole manufacturing process. Committee representation does not have to be as tight as with the design committee.

You will also have to decide who will chair the profit-sharing committee. Your choices include

- election by the committee; or
- appointment by management (human resources manager, senior management, or owner/manager).

One criterion that can be used to make this decision is the role that you determine for the committee. This is discussed in more detail in a later section of this chapter but, essentially, the committee will either be a decision-making body or an advisory group.

The advantage of having the committee elect a chair from among the members is that it promotes the concept of employee involvement. If the committee is a decision-making body, your first choice should probably be the owner/manager; the second choice would be another member of senior management. If the committee's role is more of an advisory nature, you would be more likely to choose the human resources manager or a person elected by the committee.

SELECTION OF THE COMMITTEE

The committee can either be arbitrarily appointed by management or from among those who respond to a posting. However, members could be elected by various departments. The pros and cons of either approach are the same as those discussed in Chapter 5.

TERM OF OFFICE OF COMMITTEE MEMBERS

You will also have to determine the term of office of committee members. This issue has three dimensions: length (time); degree of overlap; and dates of appointment.

The length of the term of office is a function of several factors: the frequency of calculation of the profit-sharing pool, i.e., annually, quarterly, semi-annually, etc.; the frequency of reporting of financial results; and the degree of variation in the business cycle for your particular business.

The timing of calculation of profit sharing and financial reporting, although different, are very closely connected. A related issue is the degree to which the owner/manager is prepared to disclose financial or operating data to the committee or the employees.[2] If you are unwilling to disclose much in terms of financial and operating results, meetings do not need to be as frequent.

If you decide to calculate the profit-sharing pool annually, you can start with the position that the committee should serve enough years so that reviewing the financial results will mean something to them. This suggests that a term of two or three years may be appropriate.

This does not necessarily mean that the committee only meets annually, just that it will only have to review financial results once a year. If you decide to calculate the pool more often than annually, and to share the results with the employees, you should probably have meetings at least as often. Even if the committee is meeting as often as monthly, a two- or three-year term is probably most suitable.

A related issue is whether all committee members should be elected or appointed at the same time or whether their start

---

[2] See "What to Communicate," in Chapter 11.

dates should be staggered. An example of staggering would be for one-third of the committee to be elected each year. Staggering can be useful because it ensures that at least some committee members always have experience on the committee. The obvious disadvantage is that elections must be conducted more frequently. The only main advantage of electing the whole committee at one time is that only one election is necessary when the term of office expires.

An example of staggering — assuming the committee is composed of six employees, plus the chair, and the term of office is two years — is as follows:

- initial committee appointed January 1, 1994;
- two members leave and are replaced January 1, 1995 (to serve until Dec. 31, 1996);
- two members leave and are replaced January 1, 1996 (to serve until Dec. 31, 1997); and
- two members leave and are replaced January 1, 1997 (to serve until Dec. 31, 1998).

In this example it is also possible that two new members could be elected January 1, 1994, if two of the original members declined to serve on the permanent committee. This happens quite often but does not usually pose a problem. Such employees feel that they have made their contribution and just do not feel comfortable being permanent members. This can be somewhat beneficial in that the company can get "new blood" on the committee right away.

You should also determine at the start whether committee members will be allowed to be re-elected for more than one term. In a small company this is probably desirable since it may be difficult to find enough candidates to continually have new ones every time there is an election.

There should also be provision for interim elections in case a committee member is unable or unwilling to complete his or her term. This could happen if the employee becomes disabled, transfers to a department other than the one that he or she was elected to represent or if he or she leaves the company.

TRANSITION FROM DESIGN TO PERMANENT COMMITTEE

If you have decided to have a permanent committee, the next issue is how to set it up. The first, and easiest, option is to make the initial committee the continuing body. However, this option has several disadvantages. First, members may not want to continue as committee members, even though they agreed to serve on the original design committee. It is also unrealistic to expect that committee members will continue indefinitely, even if they are willing to continue for the time being. Employee turnover is a fact of life; they may leave the company, or be transferred or promoted, and therefore may not be in the same department that they were originally selected to represent. It is also advisable to get new members periodically to ensure that the committee does not develop "paralysis" and that new ideas are considered.

Obviously, I believe that the creation of a permanent profit-sharing committee is highly desirable and that such a committee will make a significant contribution to the effectiveness of the profit-sharing plan.

RESPONSIBILITIES OF THE PROFIT-SHARING COMMITTEE

It is extremely important that the committee's responsibilities and authority be established at the outset. This will prevent confusion and disappointment later if committee members do not clearly understand what their powers are before they are elected. Possible duties of a profit-sharing committee include

- reviewing financial and operating results;
- reviewing and/or suggesting revisions to the profit-sharing plan;
- assessing employee suggestions for improving profitability;[3]
- providing general feedback to management on the operation of the plan;
- approving any proposed communications to employees about the plan;

[3] See Chapter 15, "Employee Suggestion Systems."

- reviewing investment results if the plan has a deferred component (either DPSP or EPSP);
- approving requests for withdrawals if there is a deferred component; and
- considering exceptions to the plan rules, for example, if an employee is fired but does meet the other criteria for inclusion in the plan.

ADMINISTRATION OF THE PLAN

The only major issue involving administration of the plan is defining who is responsible. If your company has a human resources manager, I believe that this is the logical person to assume this task. As discussed in the preface, I believe that profit sharing should be viewed as part of a comprehensive, progressive human resource strategy for the company. Therefore, the human resources manager, by definition, should be given the responsibility for day-to-day administration of the plan. This reasoning is even more compelling if the human resources manager is appointed to be the chair of the profit-sharing committee.

If the company does not have a human resources manager, the usual practice is to assign the job to someone in the finance function. Another option, if the chair of the profit-sharing committee is a member of senior management, but not from finance (e.g., the vice president of production) this person could be given this assignment.

REVIEWING THE PLAN

Any profit-sharing plan is designed in the context of certain circumstances that exist at the time it is designed. However, a fact of organizational life these days is that change is constant and increasingly rapid. Aside from major issues such as mergers, acquisitions and divestitures, new products and processes are constantly being introduced. Concepts such as Total Quality Management and ISO 9000 bring many innovations to existing work practices and processes. The need to downsize and reduce layers in the hierarchy also produces many changes. The economics of many businesses are being transformed because of new competition, phenomena such as the Free Trade Agreements, just-in-time manufacturing, relocation of manufacturing offshore and reduction of sales taxes (e.g., tobacco). Family businesses are sold to large conglomerates or are taken over by professional managers brought in by the family members who no longer wish to be directly involved in managing the business.

All these forces acting on the organization mean that the corporate culture in which a profit-sharing plan must operate is, in most cases, in a constant state of change. Although there are a large number of constants (e.g., the plan is used as a retirement vehicle and is therefore a DPSP), there are also a significant quantity of design factors that can be affected by changed circumstances. For example, the membership issue can be affected by the creation or purchase of a new division of the company, especially if the composition of the labour force is significantly different from existing divisions. If another company with a registered pension plan is purchased and absorbed into the parent company that is using a DPSP as the retirement vehicle, obvious revisions will have to be made. If the allocation formula uses compensation and the company makes significant changes to the pay system, such as changing to a skill-based (also known as knowledge-based) pay program or to a cash program involving significant bonuses, major changes may have to be made to the profit-sharing plan.

13
THE DESIGN PROCESS

INTRODUCTION

This chapter contains a set of checklists that the owner/manager can follow in order to ensure that he or she has covered all the elements necessary in the design of a profit-sharing plan. Simply photocopy the following pages and use them as a guide as you go through the process. If you establish a profit-sharing committee you may also want to supply the committee members with copies so they can be sure that they have covered everything.

The checklists have been organized according to the major events listed on the following page. They are placed in their logical sequence and can be used as an outline of the overall process of establishing a profit-sharing plan.

EVENT	PAGE
1. Research and Planning	133
2. First Meeting of the Profit-Sharing Committee	135
3. Committee Members' Meeting(s) with Employees	138
4. Second Meeting of the Profit-Sharing Committee	141
5. Developing and Testing of Models	144
6. Third Meeting of the Profit-Sharing Committee	145
7. Implementation	146

RESEARCH AND PLANNING

1. Initial interest and research. ( )
2. Establish the objective(s) of the plan. ( )
3. Have you chosen the type of plan you want ( )
    OR
   will you allow the employee committee input into this decision? ( )
4. Are the conditions right for a profit-sharing plan?
   - good employee relations ( )
   - competitive base salaries ( )
   - internally equitable salaries ( )
   - profits – 3-5% of base pay ( )
       – meet profit targets ( )
   - reasonable organization structure ( )
   - no other major changes ( )
   - good communications ( )
   - management commitment ( )
5. Do you intend to have a profit-sharing committee?
    Yes ( )
    No ( )
6. a. If yes, have you selected the committee members and obtained their consent to participate? ( )
   b. If yes, do the proposed committee members:
   - represent all major elements of the company ( )
   - cross section of demographics ( )
   - are opinion leaders ( )
   - are reasonably articulate ( )
   - willing to act as committee member ( )
   - capable of understanding the concepts ( )
7. Have you prepared an announcement letter for the employees? ( )
   Does it include:
   - the objective(s) of the plan ( )

- the type of plan or whether this decision will
  be left to the committee ( )
- names, titles, and departments of the
  committee members ( )
- the criteria used to choose the committee ( )
- the chair of the committee ( )
- whether the committee will make the final decision
  or just recommendations to management ( )
- the name of a person to contact if an employee
  has questions ( )
- the signature of the C.E.O. ( )

FIRST MEETING OF THE PROFIT-SHARING COMMITTEE

1. Have you scheduled the first meeting of the profit-sharing committee? ( )
   Are the necessary materials available?
   - printed background material ( )
   - refreshments, etc. ( )
   - blackboard/whiteboard/flip chart ( )
   - markers ( )
   - note taking materials for members ( )
   - other ( )
2. General
   - have all members introduce themselves ( )
   - explain the objective(s) of the plan ( )
   - cover the role of the committee members ( )
   - discuss the type of profit-sharing plan ( )
     cover the items:
       allocation of the fund:
         – length of service ( )
         – earnings ( )
         – attendance ( )
         – job levels ( )
         – equal distribution ( )
         – employee contributions (EPSP) ( )
         – merit rating ( )
         – combinations ( )
       membership:
         – minimum length of service ( )
         – as of end of fiscal period
             OR
             actual date ( )
         – cap on seniority ( )
         those who leave the company:
             death ( )

- retirement ( )
- sick leave ( )
- quits ( )
- terminations by company ( )
- minimum length of service ( )
- lay-offs ( )
- leaves of absence ( )

category of employment:
- full-time ( )
- part-time ( )
- temporary/occasional ( )
- union ( )
- commission sales people ( )
- other ( )

communications:
- full disclosure ( )
- use of ratios ( )
- use of auditor ( )
- full disclosure to committee ( )

administration:
- size of committee ( )
- elected/appointed ( )
- term of office ( )

employee suggestion system:
- part of the plan? ( )
- developed now or later? ( )

- any other concerns by the members of the committee ( )

3. If the plan is to be a DPSP, an EPSP, or a combination plan use this section, otherwise go to item 4.
   - membership ( )
   - vesting ( )
   - forfeitures ( )
   - investment policy ( )

- investment choices for the employee ( )
    - on allocation ( )
    - other ( )
- administration of invested funds ( )
- employer contributions ( )
- excess contributions ( )
- choices on leaving the plan ( )
4. General committee business
    - questions/clarification ( )
    - date for next meeting ( )
    - other business ( )

COMMITTEE MEMBERS' MEETING(S) WITH EMPLOYEES

COMMITTEE MEMBER _____

GROUP _____

DATE/LOCATION _____

1. General
    - explain the objective(s) of the plan ( )
    - cover the role of the committee members ( )
    - discuss the type of profit-sharing plan ( )
2. Plan Design
    cover the items:
        allocation of the fund:
            – length of service ( )
            – earnings ( )
            – attendance ( )
            – job levels ( )
            – equal distribution ( )
            – employee contributions (EPSP) ( )
            – merit rating ( )
            – combinations ( )
        membership:
            – minimum length of service ( )
            – as of end of fiscal period
                OR
                actual date ( )
            – cap on seniority ( )
        those who leave the company:
            –death ( )
            –retirement ( )
            –sick leave ( )
            –quits ( )
            –terminations by company ( )
            –minimum length of service ( )
            –lay-offs ( )

CHAPTER 13 — THE DESIGN PROCESS

        – leaves of absence ( )
    category of employment:
        – full-time ( )
        – part-time ( )
        – temporary/occasional ( )
        – union ( )
        – commission sales people ( )
        – other ( )
    communications:
        – full disclosure ( )
        – use of ratios ( )
        – use of auditor ( )
        – full disclosure to committee ( )
    employee suggestion system:
        – part of the plan? ( )
        – developed now or later ( )
    administration:
        – size of committee ( )
        – elected/appointed ( )
        – term of office ( )
- any other concerns of the group ( )

3. If the plan is to be a DPSP, an EPSP, or a combination plan use this section, otherwise go to item 4.
   - membership ( )
   - vesting ( )
   - forfeitures ( )
   - investment policy ( )
   - investment choices for the employee ( )
         on allocation ( )
         other ( )
   - administration of invested funds ( )
   - employer contributions ( )
   - excess contributions ( )
   - choices on leaving the plan ( )

4. General committee business
   - questions/clarification ( )
   - date/reason for next meeting ( )
   - other business ( )

SECOND MEETING OF THE PROFIT-SHARING COMMITTEE

1. For the second meeting of the profit-sharing committee are the necessary materials available?
   - printed background material ( )
   - refreshments, etc. ( )
   - blackboard/whiteboard/flip chart ( )
   - markers ( )
   - note taking materials for members ( )
   - other ( )
2. General
   a. Obtain general feedback on employee reaction to the idea of a plan ( )
   b. Have each member of the committee report on the response of his/her group to the following (this can be done by member or by item):
      allocation of the fund:
      – length of service ( )
      – earnings ( )
      – attendance ( )
      – job levels ( )
      – equal distribution ( )
      – employee contributions (EPSP) ( )
      – merit rating ( )
      – combinations ( )
      membership:
      – minimum length of service ( )
      – as of end of fiscal period
         OR
         actual date ( )
      – cap on seniority ( )
      those who leave the company:
      – death ( )
      – retirement ( )

　　　　　　　– sick leave　　　　　　　　　　　( )
　　　　　　　– quits　　　　　　　　　　　　　( )
　　　　　　　– terminations by company　　　　( )
　　　　　　　– minimum length of service　　　 ( )
　　　　　　　– lay-offs　　　　　　　　　　　　( )
　　　　　　　– leaves of absence　　　　　　　 ( )
　　　　　category of employment:
　　　　　　　–full-time　　　　　　　　　　　　( )
　　　　　　　–part-time　　　　　　　　　　　　( )
　　　　　　　–temporary/occasional　　　　　　 ( )
　　　　　　　–union　　　　　　　　　　　　　　( )
　　　　　　　–commission sales people　　　　　( )
　　　　　　　–other　　　　　　　　　　　　　　( )
　　　　　communications:
　　　　　　　–full disclosure　　　　　　　　　( )
　　　　　　　–use of ratios　　　　　　　　　　( )
　　　　　　　–use of auditor　　　　　　　　　 ( )
　　　　　　　–full disclosure to committee　　 ( )
　　　　　employee suggestion system:
　　　　　　　–part of the plan?　　　　　　　　( )
　　　　　　　–developed now or later　　　　　 ( )
　　　　　administration:
　　　　　　　–size of committee　　　　　　　　( )
　　　　　　　–elected/appointed　　　　　　　　( )
　　　　　　　–term of office　　　　　　　　　 ( )
　　• any other concerns by the members of the committee ( )
3. If the plan is to be a DPSP, an EPSP, or a combination plan use this section, otherwise go to item 4.
　　• membership　　　　　　　　　　　　　　　　( )
　　• vesting　　　　　　　　　　　　　　　　　 ( )
　　• forfeitures　　　　　　　　　　　　　　　 ( )
　　• investment policy　　　　　　　　　　　　 ( )
　　• investment choices for the employee　　　 ( )
　　　　　　　on allocation　　　　　　　　　　 ( )

other	( )
• administration of invested funds	( )
• employer contributions	( )
• excess contributions	( )
• choices on leaving the plan	( )

4. General committee business
   - questions/clarifications ( )
   - date for next meeting ( )
   - other business ( )

DEVELOPING AND TESTING OF MODELS

1. Have you developed two or three models that agree with the feedback given by the committee members? ( )
2. Have you developed a computer simulation of each of the models mentioned in the last question? ( )
3. Have you tested these models using historical, financial, and operating data for the last 3 to 5 years and for the next few years using your financial forecasts? ( )

    This exercise should include various levels of:
    - profits ( )
    - revenue ( )
    - numbers of employees ( )
    - employee compensation (base and total) ( )

4. Have you revised the models based on the above exercise? ( )
5. Have you selected the "best" option? ( )
6. Have you decided what information in the model selected in the last step will be shown to the profit-sharing committee? ( )

THIRD MEETING OF THE PROFIT-SHARING COMMITTEE

1. Have you provided a summary of the results of the second meeting? ( )
2. Have you provided copies of the models you developed? ( )
3. Did you discuss whether the models meet the feedback provided by employees? ( )
4. Did you pick the model that best meets the feedback from the employees, (best is used because there is no perfect profit-sharing plan) and review it with the committee? ( )
5. Decide whether to put the plan to all the employees for general approval or just use the blessing of the committee (subject to management final approval). ( )
6. Is there any other business? ( )

IMPLEMENTATION

1. Have you written up the plan in a formal document? ( )
2. Announce the profit-sharing plan to the company. ( )
3. Install an employee suggestion system. ( )
4. Have the first elections(or appointments) for the profit-sharing committee. ( )
5. If the plan is wholly or partly an EPSP have you established a trust? ( )
6. If the plan is wholly or partly a DPSP have you:
   - established a trust ( )
   - arranged for the plan to be registered with Revenue Canada? ( )
7. Appoint the employee who will be responsible for maintaining the program. ( )
8. Arranged a series of employee meetings in order to present the plan and answer questions. ( )

14
TRAINING OF EMPLOYEES

This chapter discusses training for employees once a profit-sharing plan has been installed. We will not address training employees to do their jobs, although the importance of that greatly increases once a plan is started. However, any training that is needed after installation was probably needed before the plan.

This chapter addresses the types of training that are necessary for an employee to be an effective member of a profit-sharing plan. This issue will be examined from two points of view: training that should be provided to all members of the profit-sharing plan and training that only applies to managerial and supervisory employees.

GENERAL TRAINING (ALL EMPLOYEES)
MECHANICS—HOW THE PLAN WORKS

The most important training for members of a profit-sharing plan concerns the mechanics of how the plan actually works. If employees don't understand the inner workings of the plan — especially how they are personally affected by it — they will not be able to relate to it and the motivational aspects of profit sharing will be lost. This topic is covered extensively in Chapter 11, "Communications." Most employers with profit-sharing plans have some formal type of communications but I have found that most, if not all, could be improved. In general, you can assume that while it is very difficult to communicate too much, lack of communication can be a major problem.

Finance

Another training need for members of profit-sharing plans is in the area of finance — that is, finance in plain language for the average employee. Financial training encompasses many areas. The first is the ability to read and understand the following financial statements:

- balance sheets
- income statements
- statements of changes in financial position
- cash flow reports

Obviously, the balance sheet and the income statement are the most important. Do not assume that this type of training applies only to rank-and-file employees. Many managerial and professional staff, outside the finance function, no more understand financial matters than the factory or office employees. Such training should include important concepts such as depreciation, extraordinary gains and losses, and cash flow. The need to reinvest profits to enhance the long-term viability of the company should also form a large part of this training. All of these elements can have a major influence on the size of the profit-sharing pool. Unfortunately, they are widely misunderstood and therefore training in this subject should be a very high priority.

Another aspect of financial training is the treatment of the salary, benefits and perquisites provided to the owner/manager of a private company.[1] Many employees will notice when the owner buys a new company car or other expensive benefit or

---

[1] With the new, executive compensation disclosure rules declared by the Ontario Securities Commission this is essentially a non-issue for public companies that trade on the Toronto Stock Exchange.

perquisite. Members of the profit-sharing plan will view such actions as reducing the size of the profit pool for no good reason. They will argue that the owner can travel just as effectively in a Chevrolet as in a Jaguar. The way to address this issue is to set aside a fixed percentage of revenues as compensation for the chief executive. For example, if the gross revenue of the company is $10 million, the owner/manager could receive 5 per cent of this amount, i.e., $200,000. These monies could be taken as salary, benefits or perquisites of any kind. This percentage would normally be consistent from year to year. How it is spent is not disclosed to employees; it is at the discretion of the owner/manager. If you choose this option, this aspect of the financial affairs of the company should be included in the training program.

INVESTMENTS

If the plan is, in whole or in part, a deferred type of plan such as a DPSP or an EPSP and it provides various types of investments that the employee can choose among in which to deposit his or her share of the profits, the employer should provide the employees with training on investment choices (see Chapters 9 and 10 on investment choices for the employee). Since most employees have little knowledge of investments, aside from interest-bearing securities, the employer has an obligation to ensure that they can make proper decisions about how their monies are invested. The program should include sections on

- retirement planning;
- when and how they can make choices/changes in their deferred portfolio;
- how to track the performance of their chosen fund;
- other investments outside the profit-sharing plan; and
- the advantages and disadvantages of the various choices within the plan.

Investment training is a very important aspect of a deferred type of profit-sharing plan. Unfortunately, it is not provided by the vast majority of employers with profit-sharing plans. The decision to provide such training goes to the very heart of the employer's responsibility to the employee. I believe that the employer would be irresponsible to provide a variety of investment options to employees without educating them in how to make those choices. It also wastes an opportunity to obtain employee goodwill. If employees do not understand the advantages of the choices made available to them, they are unlikely to credit their employer for providing them. There is also the grave danger that, without guidance, employees will make poor decisions that will not be discovered until they remove the funds. Since this will typically occur at retirement, any damage is irreversible.

CREATIVE THINKING

Another potential area of training employees is creative thinking. This aspect of training assumes that senior management is willing or can learn to listen to their staff when they suggest ideas for improving productivity. In fact, this may be one of the most important subjects for members of a profit-sharing plan. I believe that the effectiveness of profit sharing derives largely from small contributions from many employees rather than a single large contribution by a small group of individuals. Training in creative thinking will contribute enormously to achieving this goal.

LANGUAGE TRAINING

Another important area of training in Canada in the 1990s is English as a Second Language (ESL). In major urban areas such as Montreal, Toronto and Vancouver, companies typically have large number of employees whose first language is not English. Since communications is critical to the success of a profit-sharing plan, it is absolutely vital that employees be able to communicate effectively in English. (Of course, in Quebec, and possibly New Brunswick, the issue is FSL — French as a Second Language.)

EMPLOYEE SUGGESTION SYSTEMS

A further area of training concerns the employee suggestion system, assuming that you decide to provide one. Designing such a system is discussed in the following chapter. Obviously, employees should be taught how the suggestion system works but training should go beyond that. To be effective providers of suggestions, employees need to be able to

- express themselves effectively both verbally and in writing;
- calculate possible savings;
- think creatively; and
- analyse production and other processes to identify opportunities for improvement.

Although most of these items are discussed in other parts of this section, it may be appropriate to review them when training staff about the suggestion program.

Interpersonal Skills

Another important component of training employees in a profit-sharing plan should involve the building of interpersonal relationships and supporting skills such as risk-taking, constructive criticism, listening effectively and recognizing the contribution of others.

Performance Measures

The final area for training employees concerns operational statistics for their own department. This training probably does not apply to managerial/supervisory employees, at least in terms of their own area. This training should address any measures of group performance that are not corporate financial numbers, such as the following:

- tons of ore mined per hour or shift in a mining company;
- sales per square foot in a supermarket;
- cases of product shipped per day in a warehouse;
- average hours billed per consultant in a consulting firm;
- student-teacher ratios in a school board; or
- cost per employee recruited by a human resources department.

By learning about and using these numbers, employees gain a better grasp on how well they are doing. These numbers can be used to identify improvement opportunities and even to compare their section with corresponding sections in other companies.

Such performance measures are also more meaningful to rank-and-file employees than corporate financial statements, although these obviously are important as well.

Basic statistics training should also be provided to all employees who have not received formal training on the subject. This should include primary items such as averages, medians, quartiles, deciles, sampling and regression analysis. Statistics training will make employees much more effective in their use and understanding of performance measures.

MANAGERIAL/SUPERVISORY EMPLOYEES

This section outlines training recommended for employees who have managerial or supervisory responsibilities at any level in the organization. Such training could also be provided to "lead hands."[2]

ADMINISTRATION OF BASE PAY

Managerial staff should receive extensive training in administering base pay, especially if the base salary program includes a merit pay component.[3] Employees generally do not separate their opinion on pay policy (i.e., the level of external competitiveness) from the way base pay, especially merit increases, are administered. If employees believe that base pay is not administered fairly, their perception of all other forms of pay, including the profit-sharing plan, will be affected. The degree of authority for administering base pay that is given to first- and second-level management varies widely. However, you should ensure that your managers are thoroughly coached in the level of authority you have chosen to give them.

[2] For those not familiar with the term, lead hands are employees who direct other employees while doing the same type of work. For example, there are lead-hand electricians who direct other electricians. Lead hands do not hire, fire or discipline employees; such duties are reserved for supervisors and managers. In union shops, lead hands are usually included in the bargaining unit.

[3] See Chapter 16 for a review of how increases to the base salary on the basis of merit are related to profit-sharing plans.

Coaching Employees

Coaching employees who are not performing satisfactorily is an important skill in which training should be provided to supervisory staff. Supervisors must have the skills to bring employees to a satisfactory level of job performance, which is one of their major responsibilities. The most common mistake in this area is assuming that poor job performance is the employee's fault. In fact, studies have shown that the vast majority of the factors that negatively affect the behaviour of employees are under management control. For further information on this topic, please refer to an excellent book by Ferdinand F. Fournies, *Coaching for Improved Work Performance* (Bridgewater, NJ: Fournies & Associates Inc., 1978). Supervisors should know not only how to recognize the conditions that foster poor performance, but also how to correct them.

Teams

Supervisors and managers should also be trained in the area of building effective work teams. This is very important since profit-sharing plans place more emphasis on group performance than individual performance in traditional merit increase programs.

Performance Management

Management staff should also be trained in performance management, which includes

- setting and measuring objectives;
- differentiating between input and output (process versus results);
- assessing performance; and
- conducting performance appraisals.

This training should be provided even if the allocation formula does not include individual merit ratings. In fact, every company should have a performance appraisal system, although it need not be connected (see Chapter 16 for further information) to the profit-sharing plan or to any other aspect of compensation.

Managing Diversity

Another area of training for managers and supervisors is in diversity management, which is also known as employment equity. This is especially true in larger metropolitan areas of Canada where there is a diverse ethnic mix. Since profit-sharing plans promote teamwork, supervisory staff must know how to encourage employees — who often come from widely divergent cultures and sets of values — to work together harmoniously.

Other Considerations

1. If you really want training to be effective, especially training provided by managers directly to employees, reward managers for training and developing employees effectively. Make it part of their job description and their ongoing responsibility.

2. To encourage employees to broaden their skills and competencies, increase their base salary if they transfer to another job for developmental purposes even if that job is rated at the same or lower level than their present position.

3. Reimburse employees for tuition fees and the cost of books for courses taken outside normal working hours. I suggest that you broadly define which courses qualify for reimbursement. Although you could insist that the course be related to the employee's job, I believe that even a slight connection between the course content and the job responsibilities (current and possible future ones) should qualify the course for reimbursement. For example, the employee may be required to take several general arts courses (such as sociology or political science) in order to receive a degree directly related to his or her job, such as engineering or business administration.

4. A wide variety of sources of funding/financial support for job training are provided by federal, provincial and territorial governments in Canada. Since these programs tend to change rapidly, especially with changes in government, readers are cautioned to ensure that they are dealing with current programs. Contact the appropriate Ministry of Labour, your local college or university, or Employment and Immigration Canada for details on current programs.

15
EMPLOYEE SUGGESTION SYSTEMS

This chapter discusses the arguments for and against using employee suggestion programs as part of a profit-sharing plan. We will discuss how to design and install these programs and their prevalence in Canadian industry.

Generally, when a profit-sharing plan is installed, employees' enthusiasm about the plan expresses itself as the wish to suggest ideas to improve profitability. Unfortunately, usually no formal mechanism exists for them to do this. In fact, in many companies employees may have been discouraged from such action, either formally or informally. Before we continue further, the owner/manager with a strong personality should perhaps refer back to Chapter 4, to review the need for "good" employee relations and "management commitment" to employee involvement. The trendy term for this phenomenon is "empowerment."

WHY HAVE AN EMPLOYEE SUGGESTION SYSTEM?

What are the arguments for and against connecting employee suggestion systems to a profit-sharing plan? Certainly, many suggestion systems have been successfully established without profit sharing and many effective profit-sharing plans have been installed without formal suggestion programs. However, there are three main arguments for installing a suggestion system concurrently with a profit-sharing plan.

First, there are still many traditional, old-style, Theory X managers in companies today,[1] who will resist employee involvement, often very aggressively. Their theory of management states that managers must think and direct employees, who in turn carry out their orders without question. Although this assessment may sound critical, it is not meant to be except in one sense. This style of management worked very well historically. However, it is no longer viable in most situations in the 1990s. It assumes an environment where non-supervisory employees had a low level of education and/or were doing mostly manual work. This simply does not apply anymore — at least not in Canada and the United States. Therefore, Theory X managers will have great difficulty handling employee input to what they would view as "their" decisions. This argument is very similar to our discussion in Chapter 5 about management being committed to the process of employee involvement. Actually it extends the argument from the owner/manager through all levels of supervision, which in some companies could involve a substantial number of people. In view of this, a formal suggestion system will probably not change the Theory X manager's style, but it will *create a formal mechanism that gets around that style.* If you do not create such a mechanism, the only alternative to handling this problem is to spend enormous amounts of time and money retraining these managers. The cost-effectiveness of such programs is very limited.

Second, installing a suggestion system taps into the resources (education, training, imagination, initiative, enthusiasm) that employees can bring to the job if they are allowed to. If you want your profit-sharing plan to be more that just another way to distribute money to employees, then these resources must be drawn on.

The final reason for installing a suggestion program as part of the profit-sharing plan is that many of the more successful profit-sharing plans have an employee suggestion system. In *People,*

---

[1] Theory X and Theory Y were developed by Douglas Macgregor (Douglas Mcgregor, *The Human Side of Enterprise* (New York: McGraw-Hill, 1960)) to describe the behaviour of two types of managers. Basically, Theory X managers are very authoritarian while Theory Y managers believe employees can make a contribution beyond taking orders.

Performance, and Pay[2] the authors reported that 51 per cent of companies with gain-sharing plans have suggestion systems, while only 31 per cent of companies without gain-sharing plans have such systems.[3] They also found that companies with gain sharing have significantly higher levels of other forms of employee involvement such as team/group programs, Quality of Worklife (QWL) programs, quality circles, problem-solving teams, labour-management participation teams, and autonomous work teams.

There are also some disadvantages to suggestion systems, such as the cost of administration. Companies may need to hire extra staff to administer the program or existing staff may need to be assigned to this task and away from other tasks they are already performing. The cost of awards can also be a problem, since they are another cost that will be deducted from possible profits intended for sharing. Furthermore, employees may be offended when an employee whom they don't like receives an award. Both facets of this one argument reflect the belief that all employees in the profit-sharing plan should be making suggestions since their rewards lie in their share of the profit pie.

WHEN TO ESTABLISH AN EMPLOYEE SUGGESTION SYSTEM

The process of establishing an employee suggestion system can be very similar to the process involved in designing the profit-sharing plan. Early on, you must decide whether to include the process in the overall design of the profit-sharing plan or to introduce it after the profit-sharing plan has been implemented. I believe that it should be part of the development of the guidelines[4] for administering the profit-sharing plan after it is installed. As I have mentioned several times, the power of profit-sharing plans goes beyond the dollars involved; the power really lies in the potential of the plan to get employees motivated beyond their traditional levels of involvement. Offering them more money — unless it is a significant amount — will not have a

---

[2] Carla O'Dell and Jerry McAdams, *People, Performance and Pay* (American Productivity Center; 1987), 40.

[3] For the purposes of their study, O'Dell and McAdams included Scanlon, Improshare, custom-designed and profit-sharing plans in their definition of gain sharing. Their definition of gain sharing is different from mine — see Chapters 1 and 3.

[4] See Chapter 5, "General Principals of Designing a Plan," or Chapter 12, "Administration."

major effect on their motivation. This motivation must come from other methods in addition to the payouts from the profit-sharing plan. These methods include suggestion systems, TQM programs, quality circles, autonomous work teams, etc. Employee suggestion systems have the enormous advantage of being relatively easy to install and completely in line with the general philosophy of broad-based profit sharing. Since they also have many of the same issues, such as membership in the plan, it makes sense to include them as a design issue from the start.

The disadvantage of doing so, of course, is that the process becomes more complicated. Although both systems share similar issues, there are also several very different ones. This can make the process of design more difficult both in terms of time and complexity. Committee meetings could be very long.

A compromise would be to include the following questions in the original list of issues that is presented to the profit-sharing committee:

> Should the company develop an employee suggestion system that would give employees or groups of employees who put forward ideas that reduce costs or increase profitability an award of some nature (e.g., cash/stock, vacation, company products)?
> YES ❑   NO ❑

> If you answered yes, should this system be designed at the same time or after the profit-sharing plan is implemented?
> SAME TIME ❑   AFTER ❑

Committee members can discuss these issues with their units between the first and second meetings of the committee. If the general consensus is that a suggestion system should be developed concurrently, the committee members can then review the issues covered in the next section.

DESIGNING A SUGGESTION SYSTEM[5]

The following issues must be addressed in designing a suggestion system:

[5] This is not a book on employee suggestion systems; this chapter is included because such systems can, in my opinion, contribute significantly to the success of a profit-sharing plan.

- membership eligibility
- process of providing an idea
- who will evaluate the suggestions
- who will administer the program
- nature and size of the awards
- dealing with changes in equipment and processes
- treatment of the awards.

Membership Eligibility

An easy way to settle the issue of eligibility is to make membership requirements the same for both the suggestion and profit-sharing plans. However, both plans should be examined separately. Consider the profit-sharing plan first and then address membership in the suggestion program. Differences are justified in some areas.

The first area involves length of service. If you have a particularly long service requirement in the profit-sharing plan (e.g., three years), you could have a situation where employees with one or two years' experience have excellent ideas but cannot submit them. Everybody loses in such a situation. The same situation can apply to regular part-time staff who must wait even longer to qualify for the profit-sharing plan. On the other hand, if there is no service requirement and a large turnover of staff, you may receive a large number of low-quality suggestions by employees who will not be there when a decision is actually reached.

Similarly, using the category of employment (full-time, part-time, temporary, etc.) can be problematic. Often profit-sharing plans restrict membership to full-time employees and exclude part-timers even if there are significant numbers of them. This could waste an enormous pool of talent.

Another issue involving membership is whether to cap the length of service; this is only an issue in the allocation formula for the profit-sharing plan and has no relevance for a suggestion system.

The final issue concerning eligibility is whether employees can be rewarded for suggestions that apply in their own department.

That is, can a maintenance electrician receive an award that applies to the maintenance department? The answer is usually yes for all employees except management. Since managers are responsible for most, if not all, processes in their department and for making the department as efficient and effective as possible, they should be excluded from consideration for suggestions.

An extension of this issue is whether managers can make suggestions about other departments. That is, can an accounting manager present an idea for the maintenance department? Practice in Canadian industry is about evenly divided on this issue, although I favour allowing such practice. It is stretching the argument of "it's part of their job" to say that it applies to management staff in widely varying functional areas such as accounting and maintenance; sales and production; or quality control and human resources.

Employees should be allowed to submit suggestions as individuals or as a group. In a profit-sharing environment, however, it is probably more suitable to encourage group or team suggestions, at least indirectly.

Process

The process of actually submitting a suggestion is relatively simple. You can design a form, such as the sample shown on the following page, which employees can complete. As you can see, the form is relatively simple so that employees are not intimidated simply by the task of putting an idea on paper. It is important to remember that the high degree of multiculturalism in Canada means that large numbers of employees, especially in the large urban centres, are not fluent in either official language.

EMPLOYEE SUGGESTION FORM — XYZ COMPANY

Employee Name _____ Job Title _____

Department _____ Date _____

Please describe your suggestion in as much detail as possible.

Describe the benefits of the suggestions (These do not necessarily have to be quantitative.)

Employee Signature _____

FOR OFFICE USE ONLY

Date Received _____ SPA Initials _____

Referred for review to _____ Date _____

Acknowledgement sent on (date) _____

Notes:
_____
_____
_____
_____
_____

Place these forms in strategic places such as on bulletin boards, and in lunchrooms, supervisors' offices, and the human resources department; and, if the plan has an administrator other than a member of the human resources department, in that office as well. When employees have completed the forms, they will send them to a designated location, which will depend on who does the evaluations. This individual may be the employee's immediate supervisor or a suggestion plan administrator (SPA).[6] The SPA is not necessarily a full-time responsibility; these duties could be part of another employee's job depending on the size of the company and the volume of suggestions.

The program should also provide for whether employees can submit suggestions anonymously. When received, the suggestions should be date-stamped and an acknowledgement letter sent to the originator.

Who Does the Evaluations?

For the purposes of this section, I have assumed that there is a central depository for suggestions. The issue then becomes — who actually evaluates the idea? As usual, there are several choices:

- the SPA evaluates all suggestions;
- an evaluation committee is established;
- the SPA refers the idea to the person whom he or she considers best qualified — this could even be an external consultant;
- the supervisor/manager of the section to which the suggestion applies evaluates the suggestion; or
- combinations of the above.

The first option is usually only viable if there are enough suggestions — and the resultant savings — to justify a full-time employee. This will be unlikely when a plan is first established and therefore only becomes an option later.

An evaluation committee provides the advantage of bringing a variety of points of view to bear on the proposal. Another advantage

---

[6] This title is provided for convenience only and will be referred as the SPA hereafter.

is that it can be staffed with a cross-section of employees, much like the profit-sharing committee.[7] Wide representation on the committee helps to convince other employees that suggestions are being evaluated fairly and objectively. Another advantage is that a well-administered committee will bring a consistent approach to the proceedings that the other options may not. Even a single evaluator cannot usually achieve this since few staff members, especially in smaller companies, will have the expertise to review suggestions from all functional areas equally well. The disadvantages of an evaluation committee are common to all committees: scheduling, conflict with members' main jobs, difficulty in reaching a consensus, etc.

The third option is for the SPA to decide the best person to evaluate a particular suggestion. Although this would typically be someone within the company, this is not absolutely necessary. The program could be designed so that the SPA has a budget with which to hire external consultants in the most appropriate field if the expertise is not available internally. Since the suggestion plan is part of a profit-sharing plan, there would be some constraints, although they may be more intangible than specific. In this option even a temporary committee is a possibility, especially if the idea involves several major functions.

Another option is for the suggestion to go directly to the supervisor/manager of the area involved. This is the least desirable option for two reasons. First, as stated above, the reluctance of old-style managers to involve employees in decision-making may be an obstacle. They may also be reluctant to approve a suggestion because they may view it as an admission that they failed to identify a potential area of improvement. To ensure that department heads actually evaluate suggestions referred to them by the SPA, include this responsibility in their job descriptions. The second argument against this option is that employees in the department may be reluctant to present ideas since they may feel that they will be seen as criticizing their supervisor.

The advantage of this option, however, is that the supervisor or manager is probably the person most qualified to evaluate a suggestion for his or her department.

The last option is combinations. For example, the SPA could refer all suggestions to the head of the department involved but his

---

[7] See Chapter 12 on the duties of the profit-sharing committee.

or her evaluation would then have to be reviewed and/or approved by an evaluation committee. This combination has the advantage of the supervisor being involved while maintaining the advantages of the committee, especially the range of viewpoints. It also avoids the possibility of the supervisor resisting any suggestion approved by the committee without any input from him or her.

NATURE AND SIZE OF AWARDS

A number of policy issues must be addressed regarding the awards for suggestions that are accepted, aside from the most obvious decision — the size of the award. The most common practice is to pay a percentage of the amount of savings that would result from implementing the suggestion. In most plans, this percentage is based on the savings over a period of one to five years.[8] Few, if any, plans leave this calculation open-ended. Most companies pay from 5 to 20 per cent of the savings.[9]

In designing the program you must also decide whether you will pay only actual, documented savings or estimated savings. The former option obviously requires that payment be made later, which is undesirable in terms of motivation/rewards.[10] A compromise is to pay part of the estimated award, such as 75 per cent, initially and then to pay the remainder when the calculations are verified.

Most organizations also have minimum and maximum awards. The 1991 CASS Survey[11] showed that the minimum ranged from zero to $300, with a median of $50. The maximums ranged from $150 to $50,000, with a median of $10,000. One company had no specified maximum award. Two salary surveys by the Metropolitan Toronto Board of Trade reported that the average payment for a suggestion, in both surveys, was $75.[12] There is some logic to setting up a program without a maximum. Maximums are usually set on incentive programs to guard

---

[8] Canadian Association of Suggestions Systems (CASS), *Statistical Survey* (North York, 1991), 2. This survey is privately published for the members of CASS. See Appendix 1 for information on CASS.
[9] CASS Statistical Survey, 1991, 3.
[10] See Chapter 16 for the section on theories of motivation.
[11] CASS Statistical Survey, 1991, 3.
[12] *Clerical and Information Technology Salary Surveys,* (Metropolitan Toronto Board of Trade, 1994) p. 47 and 53.

against "windfall" situations, in effect saying that the employee's suggestion was not totally responsible for the savings. The question is then, if the employee's suggestion didn't create the savings, what did? A common, and workable, response to this is to set the maximum very high.

The final issue involves whether to reward suggestions that have intangible savings or benefits. For example, some suggestions may be beneficial for the company yet their benefits cannot be quantified. This is where the minimum payment is most useful.

Rewards do not necessarily have to be of a monetary nature. Other effective rewards include public recognition in the form of acknowledgement at a company event such as the annual meeting; an article and/or photograph in the staff newsletter, letters in personnel files; or even a letter from the company president.

Another powerful way to encourage employee participation in the program is to provide each person who submits a suggestion with a small gift such as a pen, pencil, mug or T-shirt along with a thank-you letter within 24 hours of receiving the suggestion. Some companies have a draw for a prize (e.g., a VCR); all staff who have submitted a suggestion in a given month or year are eligible for the draw.

PROTECTION OF SUGGESTIONS

Another issue that must be addressed is the protection of a suggestion that has not been adopted. Arguably, an employee who makes a suggestion needs some sort of "copyright" against another employee making the same suggestion later, perhaps with a different SPA or evaluation committee, and having it approved. It is inaccurate to assume that these evaluation processes are always logical and objective. An idea that appears terrific to some may seem equally silly or irrelevant to others. A recent example of this is the famous Post-it[13] note, which at first was rejected by almost everyone but has since established a new category in office products. You may decide to protect a suggestion for one or two years or even indefinitely. In the 1991 CASS Survey, 86 per cent of the companies had periods of either one or two years.[14]

[13] Trademark 3M Company.
[14] CASS Survey, 1991, 1.

Changes in Equipment or Processes

Another question is the status of the suggestion program during the introduction of new machinery, processes or even an organizational realignment. Some companies prohibit suggestions during such a period on the grounds that the changes have been initiated by management (perhaps in response to the suggestion program) and it is too confusing to consider suggestions when a change effort is already under way. The counter argument is that management needs all the help it can get; employees are on the front line and in an excellent position to contribute to the successful installation of new processes. Both arguments are equally valid and the results of the CASS Survey show this: about one-half of the companies reporting restrict suggestions during such a period while the other half do not.[15]

---

[15] CASS Survey, 1991, 1.

16
PAY-FOR-PERFORMANCE AND PROFIT SHARING

One of the usual reasons for installing a profit-sharing plan is to motivate employees to achieve objectives such as higher productivity, lower absenteeism, and, of course, greater profitability. In other words, profit sharing is used as a form of "pay-for-performance." Unfortunately, there is no generally accepted definition of the term "pay-for-performance" and therefore a wide variety of compensation plans become lumped into this broad, ill-defined and widely misused category. Consequently, there is widespread confusion about which forms of compensation are actually related to performance and to each other.

This chapter will define the term "pay-for-performance"; discuss theories of how pay can be used for motivational purposes; show how profit sharing fits into the concept; and discuss other forms of "pay-for-performance," such as productivity gain sharing, along with their advantages and disadvantages especially compared with profit sharing.

THEORIES OF MOTIVATION

There are two main psychological theories to explain what motivates people to behave in certain ways: expectancy theory and equity theory. The issue of whether money is a motivator is continually debated but essentially involves the question of whether money is important to the individual. Although this may sound obvious, the question is really whether "more" money is important — not just money itself. This is also a question of relativity. All organizations have various rewards they can offer to employees in return for behaviour that the organization considers desirable. These include both tangible and intangible items such as recognition, benefits, perquisites, titles, office size and decoration, promotion, training opportunities, career growth, bonuses, pensions, praise from management, and career and personal growth. However, employees are also motivated by factors outside the organization such as family, health concerns, and political and social trends.

EXPECTANCY THEORY

Expectancy theory proposes that all people, in all roles, such as parent, child, spouse, church member, consumer, and, for the purposes of this book, employee, are motivated by three "expectations."

First, any reward that is offered to them in hopes of affecting their behaviour must be of value to the individual. Simply put, an employee will not be motivated by the prospect of a reward that he or she does not value. It is critical to recognize that any perception of value must be in terms of the individual employee(s). It does not matter if you, the manager, believe that the reward is important — how the employee perceives it is what matters. This perception of value has two dimensions: the nature and the amount of the prospective reward. For example, the employer might offer an employee the prospect of a promotion to a higher-level job such as from labourer to supervisor. Suppose that the difference in pay between the two jobs is 15 per cent. The employee could consider such issues as the amount of extra work involved; the fact that she will have to be the "boss" of her former buddies; that extra training might be required; that union protection may be lost; and balance these against the prospect of a 15 per cent (before-tax) raise and the remote possibility of further promotions. To determine whether these considerations will appeal to the employee, we must consider her views, not our own. Evaluating and balancing all these factors will be carried out by her, strictly in terms of her own values and needs. If she decides that the promotion is undesirable, it cannot be used to motivate her, under any circumstances, even though it may motivate other employees who have similar jobs.

Another part of the perception of value concerns the size of the actual payment, both relative to other rewards and in its own right. This applies especially to cash rewards such as bonus payments and salary increases. We already discussed the need for cash payments to be at least 3 to 5 per cent of base pay in Chapter 5.

The second "expectation" of expectancy theory is that employees must believe that the behaviour that management is requesting is within their power — in other words, that they can do what they are being asked to do.

Again, this appears straightforward but is deceptive. The sales representative who is asked to double his sales volume in the next quarter in return for a bonus of $10,000 (when his base salary is $30,000) may find the proposed reward extremely attractive but is absolutely certain that the sales target is not achievable. In this case, the reward will not motivate the sales representative.

The third component of expectancy theory is that employees must believe that management will fulfil their commitments. That is, employees must be convinced that the company will provide the promised rewards if employees reach their goals. The major elements in this equation are visibility and magnitude. Employees must see/experience the employer actually deliver on the promises. This is where so-called "merit pay" programs frequently fail since the size of, and reasons for, all merit salary increases are usually kept secret. Although the employer may claim that all salary increases are based on performance, employees cannot verify this simply because the data are not available. Promotions, on the other hand, are highly visible although they are not typically as attractive to as many people as cash bonuses or salary increases.

EQUITY THEORY

The other major theory about employee motivation is called "equity" theory. This theory proposes that all employees compare, often instinctively, the inputs that they perceive they bring to the job (such as education, experience, personal skills, attitude to work, opportunities foregone elsewhere, etc.) and the outputs they receive from the employment relationship. These outputs include both intangible and tangible rewards from working at that company, including job satisfaction, wages, salaries, other cash payments, benefits, perquisites, social needs, career opportunities, training and so on.

If the employee perceives that these inputs and outputs are not in balance, he or she will take whatever action or inaction is required to achieve a sense of equilibrium. The most obvious possibility in this case is to quit the job. Other possibilities include decreasing productivity; increasing absenteeism; more sick leave; pilfering or theft; resistance to new work methods or machinery; and avoidance of general involvement in the operations of the enterprise.

The problem is that these acts designed to create equity in the mind of the employee are usually neither obvious to nor observable by management.

DEFINITION OF PAY-FOR-PERFORMANCE

The term "pay-for-performance" is widely used in business; however, in all the literature on compensation in organizations, no general definition exists for this expression. It is a widely misused and misunderstood term. For the purposes of my management consulting practice in the field of compensation, I have developed the following definition of pay-for-performance:

> Pay-for-performance exists when at least part of the cash compensation of the employee is determined by the performance of the individual or the group or organization to which he belongs.

There are several key words in this definition: part, cash, performance, and the term "individual or group." The term "part" refers to the fact that only a portion of the compensation package can, or should, be used for motivational purposes. A large part of the package is used only to encourage the employee to join and stay with the organization (the compensation practitioner's jargon for this is "attract and retain"). This is true despite the fact that many sales representatives work on what is called "straight commission." Many people believe that all such pay is performance-based while in reality, in most cases, a significant portion of employees' compensation on straight commission is earned regardless of the employee's conduct. Thus, only "part" of the package for these individuals is performance-related.

"Cash" is also part of the definition of pay-for-performance because it has many advantages over other types of rewards, such as benefits, tenure, promotions or training opportunities, for motivational purposes. Cash has the advantages of almost universal appeal, flexibility in both timing and amount (in contrast with a pension plan, for instance) and can be used repeatedly, provided the funds are available.

The use of the term "performance" is fairly obvious in a definition of pay-for-performance but it must be linked to the phrase "individual or group." Pay-for-performance can be determined by the behaviour of either the individual or the group. The principles of expectancy and equity theory apply, albeit slightly differently, in both situations. This concept that performance can be both individual or group is probably new to many compensation practitioners, who tend to restrict the idea of pay-for-performance to merit salary increase plans that are solely related to the single employee.

TYPES OF PAY-FOR-PERFORMANCE PLANS

Given the above definition, there are three major types of pay-for-performance plans:
- salary increase plans
- individual bonus or incentive plans
- group plans

Salary increase plans use merit or performance to determine at least part of the increase in the employee's base salary. The new salary remains in place and becomes, like all other salary, a new fixed cost.

Individual bonus plans pay a cash award over and above the base salary, the size of which is determined by the employee's performance. The bonus payment is made once and is usually not repeated unless the performance is repeated. This definition includes a wide variety of pay plans including commissions, factory incentive plans, executive bonus programs, and individual piecework plans. At first glance, these compensation plans seem widely diverse but they all share the common features of cash payments, in addition to salary, that are determined by the individual's behaviour.

Group plans, on the other hand, share with individual bonus plans the feature of the cash award that is in addition to salary. The difference is that, with group plans, the size of the payment is largely determined by the performance of the group (there are some minor exceptions to this — see Chapter 7 on allocation). There are two main sub-categories among group plans: productivity gain-sharing plans and profit sharing.

Productivity Gain-Sharing Plans

Within this category are three classic gain-sharing plans: Scanlon, Rucker and Improshare. Although each plan was developed by a different person and in a different era, they all have several common features:

- the establishment of a productivity standard at a point in time;
- the periodic measurement of gains in productivity over that standard that results in the determination of a pool of money;
- the pool is divided among the members of the plan on a purely mechanistic basis, i.e., the individual payment is in proportion to the base salary; and
- payments are made monthly.

Let's look at one example of a gain-sharing plan to illustrate the principles involved. The example is from a Scanlon plan, which is the best-known and oldest of the classics. It is also greatly simplified.

EXAMPLE SCANLON PLAN CALCULATION	
Sales	$1,000,000
Less returns	50,000
Net sales	950,000
Additional inventory	30,000
Value of production	980,000
Allowed payroll costs* (25% of production)	245,000
Actual payroll costs	200,000
Bonus pool	45,000
Employee share (75%**)	33,750
Payout	33,750 / 200,000 = 16.9%

In this example each employee would receive 16.9 per cent of base pay for that period as the bonus payment.

*Allowed payroll costs are usually calculated on a historical basis. They are what I referred to above as a productivity standard.

**In classic Scanlon plans, employees share 75 per cent of the gains each month. The percentage is different for Rucker and Improshare plans.

DIFFERENCES BETWEEN PROFIT SHARING AND PRODUCTIVITY GAIN SHARING

Profit sharing differs from gain sharing in several significant ways. Obviously, profit sharing is based on profitability but it is also different in that it is calculated each fiscal period as a portion of the profits for that period. In other words, it is an absolute calculation each time. e.g., 5 per cent of net profits before tax for that period. Gain-sharing plans, on the other hand, calculate the bonus pool in terms of the difference between the standard measurement of productivity and the gain over the standard during that period. The reference point is always the historical standard.

Another difference between the two types of group plans is that productivity gain sharing (at least for the classics — Scanlon, Rucker, and Improshare) always divide the pool of money by prorating it to base salary. Profit-sharing plans also use this often in conjunction with seniority, but also use a wide variety of other methods (see Chapter 7,"Allocation: Dividing the Funds").

Profit-sharing plans also have the advantage of not requiring additional financial reporting systems; they can use established and accepted methods. Productivity gain-sharing plans are usually installed at the plant or department level rather than throughout the whole company or profit centre as are profit-sharing plans. Productivity gain-sharing plans usually pay monthly while the vast majority of profit-sharing plans pay annually.

Comparison of Types of Pay-for-Performance

What if you are considering using some form of compensation to motivate your employees to higher and/or better levels of performance? What type of system should you use? The following table compares the three major types using a number of important criteria.

CRITERIA	TYPE OF PAY-FOR-PERFORMANCE			
	MERIT INCREASES	INDIVIDUAL BONUS	GROUP PRODUCTIVITY GAIN SHARING	PROFIT SHARING
Effort to implement	HIGH	LOW	HIGH	MODERATE
Effort to administer	HIGH	LOW	HIGH	MODERATE
Needs performance appraisal system	YES	YES	NO	NO
Motivates group behaviour	NO	NO	YES	YES
Motivates individual behaviour	YES	YES	NO	NO
Motivates individual rivalry	YES	YES	NO	NO
Motivates group rivalry	NO	NO	YES	NO

CONCLUSION

Profit-sharing plans have a variety of features that enable them to be used by employers as an incentive to stimulate different types of behaviour. The fact that profit sharing is so adaptable to different situations and purposes is one of the reasons it is so useful. Both types, cash and deferred, have at least some appeal to all employees and can be used to motivate various types of performance such as improved productivity (for which cash is better) or long-term service and loyalty (for which deferred plans are preferable). Profit sharing can be combined with other types of pay-for-performance although employers must be careful to avoid over-complication.

17
PROFIT SHARING AND UNIONS/ORGANIZED LABOUR

INTRODUCTION

Can profit sharing thrive, or even exist, in a company where some employees are unionized? In short, the answer is yes. This chapter will discuss the issue of unions on the design and administration of a profit-sharing plan in various situations. It will also review the attitudes and opinions of labour organizations about profit-sharing plans for their members. To put the question of labour relations in the proper perspective, we have included a brief overview of the legislative framework for labour relations and human resources law in Canada.

CANADIAN LABOUR LAW

In Canada, responsibility for labour relations is divided between the 10 provinces and the federal government. Certain industries (such as banking, railroads or telecommunications) are federally regulated in general and therefore also considered subject to federal labour law. In all other industries, labour legislation is the prerogative of the provinces. All employees of an employer company who work in Saskatchewan, for example, are subject to the laws of that province. Although several pieces of legislation may affect the relationship between employees and their employer, typically one major act applies to union-management relations.

VIEWS OF THE LABOUR MOVEMENT

In general, the labour movement strongly and adamantly opposes profit sharing. In fact, the labour movement is opposed to almost any form of contingent or variable pay. For example, the following statement by Cheryl Kryzaniwsky, President of CAW Local 2213, was in response to a speech by Hollis Harris, president of Air Canada, in which he said that the company was examining various ways of linking wages to corporate performance, including profit sharing.

> The CAW [Canadian Auto Workers] would never take a performance-based salary because our workers don't make the management decisions that affect the performance of this company...I don't see why our salaries should be tied to bad management decisions.[1]

Since this was said in 1995, it is interesting to look back compare it with and examine the following statement made in 1916 by Samuel Gompers, president of the American Federation of Labour:

> This proposition has never been seriously considered by the organization of labour. I desire to say further that it has come under my observation that some employers who have inaugurated systems of so-called profit sharing have pared down the wages of their employees so that the combined sharing of profits and their wages did not equal the wages of employees in other companies in the same line of industry. What we are especially interested in more than profit sharing is a fair living wage, reasonable hours and fair conditions of employment.[2]

---

[1] *The Globe and Mail,* February 18, 1995, B5.
[2] Charles R. Perry and Delwyn H. Kegley, *Employee Financial Participation: An International Survey* (Philadelphia: Industrial Research Unit, The Wharton School, University of Pennsylvania; 19104-6358, 1990), 31.

The similarity between the two statements, made almost 80 years apart, is amazing. While they will probably seem fairly inflammatory to the average owner/manager, the labour movement tends to have a sound philosophical and theoretical opposition to any form of "contingent" pay, which, of course, includes profit sharing. This opposition usually includes one or more of the following four positions:

1. Participating in profit sharing would mean that labour's income will fluctuate with the prosperity and competitive position of the firm. Union leaders prefer fixed wages and benefits for their members.

2. There is no point in participating in contingent pay programs because union members are not allowed to participate in decisions that truly affect profitability. Organized labour is also concerned that the worker will be blamed for management's mistakes.

3. In many cases, unions are not involved in the internal administration of profit-sharing plans. In other words, there is little communication between the parties and the union leadership is forced to rely upon management for the little information they receive. As a result, no trust develops.

4. Profit-sharing plans are a form of labour-management cooperation and participation in them undermines the collective bargaining process that is, almost by definition, adversarial in nature.

COMPANIES WITH UNIONS AND PROFIT-SHARING PLANS

Although the labour movement largely opposes profit sharing, many companies have instituted such plans with their unions, including the following:[3]

---

[3] This list was prepared by the Bureau of Labour Information from their database of companies that are federally regulated and have 500 or more employees in their bargaining unit and provincially regulated companies with 200 or more.

COMPANY/LOCATION	UNION	DURATION OF UNION CONTRACT
Algoma Steel Sault Ste. Marie	U.S.W.A. Local 2251 and unnumbered Local	1995
Quebec Cartier Mining Company, Lac-Jeannine, Quebec	U.S.W.A.	Mar./93–Feb./96
Inco Limited Sudbury, Ontario	U.S.W.A.	Apr./92–Mar./95
Thompson, Manitoba	U.S.W.A.	Sept./93–Sept./96
Hudson Bay Mining and Smelting Co., Limited	U.S.W.A.	Oct./93–Sept./96
Waterville TG Inc. Waterville, Quebec	United Rubber, Cork, Linoleum and Plastic Workers of America	Aug./93–July/96
NBIP Forest Products Inc. (Division of Avenor Maritimes Inc.) Dalhousie, N.B.	Communications, Energy and Paperworkers Union of Canada	May/93–Apr./98
Abitibi-Price Inc., Papeterie Kénogami Kenogami, Quebec	Federation of Paper and Forest Workers	May/93–Apr./98
Spruce Falls Inc. Kapuskasing, Ontario	**	Oct./94–Sept./99
Sidbec-Dosco Inc. Contrecoeur, Quebec	U.S.W.A.	Apr./94–Jan./97
Stelco Inc. Hamilton, Ontario	U.S.W.A.	Aug./93–July/96
IPSCO Inc. Regina, Saskatchewan	U.S.W.A.	Nov./94–July/97
Cominco Ltd. Trail, British Columbia	U.S.W.A.	Oct./92–May/95

** Communications, Energy and Paperworkers Union of Canada; International Brotherhood of Electrical Workers; Office and Professional Employees International Union; IWA-Canada.

COMPANY/LOCATION	UNION	DURATION OF UNION CONTRACT
Case, J.I., Company (Division of Tenneco) Hamilton, Ontario	U.S.W.A. 2 locals	May/94–Apr./97 Oct./94–Apr./97
Transport Provost Inc. Province-wide, Quebec	Communications, Energy and Paper-Workers Union of Canada	June/93–Apr./96
CP Express & Transport Canada-wide	Transportation-Communications International Union	Sept./94–Aug./99
Société des alcools du Québec Province-wide, Quebec	Canadian Union of Public Employees	July/93–Dec./98

INSTALLING A PLAN IN A UNION SITUATION

You will face three possible scenarios if you are considering installing a profit-sharing plan in a unionized setting:

- there is one bargaining unit with some of the non-management staff excluded from the unit;
- there are two or more bargaining units with some non-management staff not included in any of them; or
- none of your employees are unionized but you have either had a recent attempt at certification or suspect that one will be attempted in the near future.

I will discuss below the major issues that are unique to each scenario. Each discussion also includes a number of other variables such as the length of time the unit(s) have been certified; the quality of the union/management relationship; the relationship between the bargaining units; and the profitability and future prospects of the business.

SITUATION 1 — ONE UNION FOR SOME OF YOUR EMPLOYEES

In this situation, two basic issues arise:

1. Should the plan include union members or just the non-union employees?
2. If you decide, or wish, to include the unionized employees in your profit-sharing plan, to what extent are you obligated to negotiate the plan with the union?

In terms of the first issue, the initial approach should be that union members should be included since you are considering a "broad-based" plan. However, some circumstances would indicate an alternative approach. One approach could be that the bargaining unit already has some sort of group incentive plan,[4] such as a Scanlon or Improshare productivity gain-sharing plan. In this situation, including the union members in the profit-sharing plan would probably reward them doubly (i.e., out of proportion to their contribution) for their performance as a group.

Another condition that would preclude inclusion of the members of the bargaining unit is simply that the leaders of the local, or even the executive of the parent union, oppose the idea.

To determine whether you need to negotiate the profit-sharing plan with the union, you should examine the recognition clause in the collective agreement. The recognition clause is usually the first clause in the agreement. It states that the company (or employer) recognizes the union as the exclusive bargaining agent for a specified group of employees (e.g., all production employees below the rank of supervisor or foreman) on specific subjects such as hours of work, lay-offs, wages and benefits, vacations, and all types of premium pay. Check your collective agreement to see if it refers to profit sharing or any type of incentive pay in the recognition clause.

If your collective agreement refers specifically to profit sharing, there is little doubt that you will have to negotiate your plan. If the recognition clause does not mention profit sharing or incentive plans in general, you should refer to the relevant case law, which presents a divided picture that will, of course, vary according to the labour law jurisdiction in which the bargaining unit is located.

---

[4] See Chapter 16 for a discussion of different types of group pay-for-performance plans.

CHAPTER 17 — PROFIT SHARING AND UNIONS/ORGANIZED LABOUR

Arbitration cases on the subject of incentive rates also vary as to whether employers can introduce them without consultation with the union. Brown and Beatty say:

> Absent some particular clause in the agreement to the contrary, the establishment of such a method of remuneration is generally assumed to fall within the management prerogative....However, arbitrators have differed as to management's right to unilaterally introduce a wage incentive system, in the absence of some express provision in the agreement, where none had existed before.[5]

Again, the use of an industrial relations consultant or labour lawyer is recommended.

SITUATION 2 — TWO OR MORE BARGAINING UNITS

If you have two or more bargaining units in the potential profit-sharing plan, the issues outlined above under Situation 1 are the same except they are multiplied by the number of bargaining units.

However, with multiple bargaining units, there is an additional problem: the challenge of getting various unions to agree with each other, in addition to agreeing with you. It is almost impossible to provide you with guidelines or instructions on how to proceed in this situation except to alert you to the following issues:

- having the same rules for all units is a highly desirable objective;
- however, it is possible to have different contribution, allocation and membership rules for each bargaining unit;
- having different rules will greatly complicate the design process and the need for a sophisticated and extensive communications program will be extreme; and
- different rules may make the process of transferring employees from one unit to another, or from a union job to a non-union one, very difficult. For example, they could find themselves receiving a smaller share of profits by changing units.

---

[5] Donald J.M. Brown and David M. Beatty, *Canadian Labour Arbitration*, 3rd ed. (Aurora, Ontario: Canada Law Book Inc.), Release No.1, 14; December 1994, 8–22 to 8–22.1.

SITUATION 3 — NO UNIONS

If you do not have any unionized employees, you can basically proceed to develop a profit-sharing plan according to the procedures described in this book. However, I caution you to avoid using the profit-sharing plan as a union-avoidance device. First, this strategy is highly unlikely to work. Employees will quickly see through this strategy. Be sure to review Chapter 4, "Conditions Needed for Profit Sharing," especially the sections about communications, trust and management commitment before proceeding. If a union has initiated certification proceedings against you for any of your employees, you will probably contravene the relevant labour legislation by trying to initiate a profit-sharing plan at the same time. I recommend that you consult a labour relations consultant or lawyer in such a situation.

TRENDS IN UNION SHOPS AND PROFIT SHARING

Is there any pattern to the design of profit-sharing plans in companies with unionized employees?

To begin answering this question, you can look at Chapter 19, "Sample Profit-Sharing Plans," in which there are three companies that include unionized employees in their profit-sharing plan. There are also the companies listed on pages 176–177.

For an in-depth analysis, see Professor Anthony E. Smith's 1985 study of nearly 1000 firms across key sectors of the Canadian economy, in which he asked about workplace innovations introduced between 1980 and 1989 in the major bargaining units of the companies contacted. He found that 24 per cent of the companies had introduced profit sharing in this period. The only other workplace innovations that were more prevalent than profit sharing were flexible work systems (29 per cent of companies) and labour–management committees (32 per cent of companies).[6] The fact that these two innovations were also very popular in union settings indicates a positive future for profit sharing as both are the type of changes that are very supportive of, and compatible with, profit-sharing plans.

---

[6] Anthony E. Smith, "Canadian Industrial Relations in Transition," *Industrial Relations*, Vol. 40, No. 4, 1985. 651.

CHAPTER 17 — PROFIT SHARING AND UNIONS/ORGANIZED LABOUR

In addition to the evidence of the many companies listed above having experience with unions and profit sharing, it appears possible that the opinion of big labour may be changing as well. The following quotation from an AFL-CIO report — although cautious — encourages some optimism:

> Labor unions are in favor of improving productivity, increasing job security, and having a healthy and growing economy. However, unions have yet to see any convincing arguments that profit sharing is a way to achieve these goals. Even with these doubts, most unions are willing to negotiate with employers about creating a profit-sharing plan as a benefit that is in addition to the basic wage and benefit package.[7]

---

[7] AFL-CIO Report No. 49, May, 1991, 2.

18
PROFIT SHARING IN THE UNITED STATES AND MEXICO

INTRODUCTION

The North American Free Trade Agreement (signed in December 1993) created a free-trade zone including Canada, the United States and Mexico. Since many Canadian companies considering a profit-sharing plan will have, or may start, foreign subsidiaries, I believe that it is appropriate to briefly review profit sharing in the United States and Mexico.

It is important for Canadian companies to understand the American and Mexican situations for several reasons. The most obvious reason is that you may decide to establish operations in either country. If you have a broad-based profit-sharing plan in Canada, it is quite possible that you could extend the plan to include your Mexican and American employees. Or you may decide to set up separate plans for the operations in each country.

Another reason is that Mexican or American companies with profit-sharing plans may establish subsidiaries in Canada that could compete with you directly, in your line of business, or locate nearby and compete for the same types of skilled labour that you need. It is quite obvious that you have to be competitive with the companies that you compete with for labour, such as the steel mill that employs the same types of electricians and computer operators that you do. However, you also have to have labour costs that are comparable to companies in the same business that you are in. The existence of a profit-sharing plan may give other companies a cost advantage. In Chapter 5, where we reviewed the different reasons why employers install profit-sharing plans, one reason was to reduce fixed costs such as wages and salaries. If your competitors are using profit sharing for this reason, you may be at a serious cost disadvantage.

This chapter will address profit sharing in both countries as well as the question of whether employees in other countries should be included in the Canadian plan or in a separate plan for employees in either Mexico or the United States.

THE UNITED STATES

In the United States, labour law is regulated by the federal government, unlike Canada where it is primarily the obligation of the provincial governments. However, as in Canada, where the primary tax legislation is the federal *Income Tax Act*, in the United States it is the federal Internal Revenue Code.

There are three types of profit-sharing plans in the United States that are very similar to the types available in Canada. As in Canada, many companies have cash profit-sharing plans. The monies are paid directly to the employees and taxes are withheld at source. All funds paid out are deductible to the employer and there are no legislated limits to the amount paid.

The other two types of profit sharing in the United States are deferred profit-sharing plans and 401(k) plans. The latter are a variation on the former. The alpha-numeric designation refers to the relevant section of the Internal Revenue Code.

With deferred plans, money is placed into a trust, which is invested and the money is normally taken out at retirement. The monies are not taxable to the employee until they are removed. The 401(k) plans were created by a revision of the Internal Revenue Code in 1978 and their growth has been spectacular since then. These plans are established by adding the 401(k) provision to a "qualified" retirement plan. Qualified plans include three types: profit-sharing plans, stock bonus plans and a certain type of money purchase plan. These 401(k) ( ) plans are closer to the Canadian DPSPs than what Americans call deferred profit-sharing plans (apparently, Americans do not use a similar acronym). However, there are also some major differences.

In 401(k) plans the employee may defer up to $7,000 of pre-tax income. The employer may contribute up to $30,000 (which includes the $7,000) on behalf of the employee and treat the amounts as deductible up to 15 per cent of payroll. The income earned within the plan is not taxed until it is withdrawn. An owner of a business is considered an employee for the purposes of 401(k) plans. This is in dramatic contrast with DPSPs in Canada. Although the employer contributions to 401(k) are often based on profit levels, they are also frequently based on employee contributions. Employees can also get loans from a 401(k) plan.

It is difficult to obtain the exact number of profit-sharing plans in the United States. However, surveys indicate what I consider to be two clear trends: profit sharing is probably more prevalent in the United States than in Canada and 401(k) plans are much more popular than cash plans. The latter observation appears to be the opposite of the situation in Canada.

One survey in 1994 of 1000 employers indicated that 25 per cent offered a profit-sharing plan while 56 per cent provided a 401(k) plan.[1] In 1989, a Bureau of Labor Statistics survey showed that, in medium and large firms, 16 per cent of employees were included in profit-sharing plans. Of these, 1 per cent were in cash plans; 13 per cent were in deferred plans; and the remaining 2 per cent were in combinations.

Discretionary employer contributions seem to be more prevalent in the United States than in Canada. According to the Bureau of Labor Statistics, 40 per cent of plans do not have a predetermined formula. The same survey indicated that the most dominant form of allocation formula (64 per cent) was earnings by itself.

MEXICO

Unlike the other two countries in NAFTA, Mexico takes a different approach to profit sharing. In Canada and the United States, profit

---

[1] Martha Priddy Patterson, "1994 KPMG Retirement Benefits Survey," *Journal of Compensation and Benefits*, March–April, 1995, 6.

sharing, although subject to legislation, is voluntary on the part of employers. They do not have to have a profit-sharing plan.

In Mexico, however, profit sharing is mandatory for all employers with over 10 employees (there are some exceptions including new companies for the first year of operations). In fact, it was made part of the country's constitution in 1916–17 but did not become really effective until the early 1960s when the constitution was further amended.[2]

Mexico is a federal state like both Canada and the United States. It has 31 states and a federal district, which is basically Mexico City, and has tax jurisdiction similar to both a state and a city. The federal and state governments have separate areas of legislative jurisdiction but labour law is the responsibility of the federal government. This is similar to the split in the United States but different from Canada where the provinces are responsible for labour legislation except for companies in federally regulated industries. Mexico has one statute governing labour relations — the Federal Labour Law.

The legislation that requires mandatory profit sharing is found in the Mexican constitution, the Federal Labour Law and the country's income tax law. Employers are required to share 10 per cent of their pre-tax profits with all employees with some minor exceptions.[3] These exceptions include the general manager and any temporary employees who have worked less than 60 days during the year. Companies that are in their first year of operation are also exempt. Although profit-sharing payments are deductible to the employer, you should note that, in calculating pre-tax profits, employers cannot deduct loss carryforwards or some other items.

The 10 per cent of profits is divided among eligible employees in two ways. First, the 10 per cent is divided in half. The first half is distributed in proportion to the days or hours worked by the employees — in other words, by attendance (see Chapter 7's section on attendance). The other half of the total pool is paid in proportion to the base salary (see the section on earnings in Chapter 7) earned by each employee during the year. The highest salary that can be used for this purpose cannot be more than 20

---

[2] Susan Purcell, *The Mexican Profit Sharing Decision* (Berkeley: University of California Press, 1975), 47–48.

[3] This 10 per cent is supposed to be reviewed every 10 years and could be revised. Past revisions have all been upward.

per cent higher than the highest-paid union members. Basically, the Mexican mandatory profit-sharing plan has a combination allocation formula that includes attendance and earnings.

Apparently no companies in Mexico offer a "voluntary" profit-sharing plan to their employees that is over and above the legal requirements.

SEPARATE OR UNIVERSAL PLANS

If you have a broad-based employee profit-sharing plan in your Canadian company and establish operations in either the United States or Mexico, you will have to consider whether to establish a profit-sharing plan for those operations.

This issue is a variation on the question of membership, which is examined in Chapter 6. The essential question is whether you should include employees in either Mexico or the United States in your Canadian plan, using the same membership criteria, or whether you should implement separate plans for your foreign operations.

First, you should examine whether to include your foreign employees in the Canadian profit-sharing plan. The first issue to look at is the degree of autonomy of the foreign operations — that is, are they true, separate profit centres? If they are not real profit centres, you would be inclined to include them in the Canadian plan. On the other hand, if they are actual profit centres, having their own profit-sharing plan will probably be more motivational for obvious reasons.

Another issue is the objectives of the plan(s) and the type of plan that this objective requires. The type of plan permitted by the laws of the respective countries will also influence this subject. For example, if the objective of the Canadian plan was to provide retirement income, you will have chosen a DPSP or, less likely, an EPSP. This means that, if you wish to achieve the same objective, you will need a vehicle similar to DPSPs. As you have seen from the discussion earlier in this chapter, the United States has both deferred profit-sharing plans and 401(k) plans. However, Mexico does not have such tax-sheltered vehicles. Cash plans are not as difficult in this area as all three countries allow cash profit sharing and do not restrict the amounts that the employer can deduct.

Other factors to consider in this decision include
- the demographics of your employees;
- the level of earnings;
- other benefits that are legally required (for example, in Mexico employers are required to pay 15 days' pay as a Christmas; and
- the degree of profitability and level of maturity of the operation (is it a start-up situation or a smooth-running organization that has been going for many years?).

If you decide that you need to have separate plans for a foreign subsidiary, the first sub-issue to address is whether to transfer the Canadian plan in its entirety or to revise it to meet local conditions. You should carefully examine the three main aspects of your plan to see if they can be transferred.

The first of these is the employer contribution formula, i.e., how much money the company puts into the profit-sharing fund. As pointed out in Chapter 5, this formula is developed from two points of view: as a percentage of the earnings of the plan members and as a percentage of profits either before or after taxes. Since both of these are likely to be different from the Canadian situation, it is quite probable that the employer contribution will have to be different. Of course, this could be as simple as just being a different percentage of net profits before tax.

The second major feature is the membership criteria (see Chapter 6). The probability that this can be transferred to the new country without change is higher than for the employer contribution formula. However, the following factors could affect this:
- age restrictions (e.g., in the U.S., employees can be excluded from a 401(k) plan if they have not reached the age of 21);
- the use of staff who are called "contractors" but in Canada would be considered full-time employees because they only work for the company;
- union status (the degree of unionization is much higher in Mexico and much lower in the U.S. than in Canada); and
- other legal requirements.

The final major feature of your plan that you should review is the allocation formula. Because of the similarity between the business cultures of Canada and the United States, it is very probable that you could use the same formula in both countries. However, Mexico presents at least two variables that could have a major impact that you should examine carefully.

First, the business culture is quite different in several ways. For example, it takes two to three times longer in Mexico to get things done. In addition, Mexican managers take a more paternalistic approach to decision-making than Canadians or Americans. Mexican workers and technical or administrative staff are very reluctant to criticize more senior members of the company. They are also much more reluctant to go beyond the specific instructions received. It is difficult to encourage decision-making at lower levels.

Second, the compulsory profit-sharing laws already specify an allocation formula, as discussed above, that includes both attendance and earnings. If you introduce a plan with a different allocation formula, there are two dangers. First, the difficulty of communications increases dramatically simply because of the increased complexity. Second, the new allocation rules may be in conflict with the legislated rules, not in legal terms, but in terms of motivation. For example, if you include a merit component, it could both be in conflict with the legislated formula and be incompatible with the general aspects of Mexican culture, which put more emphasis on harmony of the group rather than individual concerns.

CONCLUSION

Canadian companies with operations in the United States would have relatively little difficulty establishing profit-sharing plans for their American employees that would be very similar to their Canadian plans, if not exact duplicates. Although this is not true in Mexico, I believe that it is possible. However, the virtual non-existence of profit-sharing plans above the legal minimum probably makes this a non-issue.

19
SAMPLE PROFIT-SHARING PLANS

This chapter contains 12 summaries of profit-sharing plans of all types covering a representative cross-section of Canadian businesses. Factors such as industrial sector; location and size (number of employees and sales volume); date the plan was installed; union status; length of time in business; and type of plan were all used to select the employers who were asked to provide this information.

The data shown were collected in late 1994 and early 1995. Special thanks are due to the companies and executives who provided the information and granted permission to include these data in this book. The person who provided the information is listed for each plan.

These sample plans are simply included to give you an idea of the variety and range of possibilities available as you start to design and implement your own profit-sharing plan. Remember that there is no "right" formula for a profit-sharing plan. Each plan must be designed in the context of the objectives set by you, the owner/manager, at the beginning of the process (see Chapter 5 on the importance of clearly establishing your objective) and the culture and values of your company and its employees.

COMPANY	LOCATION
A & G Hodgkinson Sales Ltd.	Barrie, Ontario
Algoma Steel Inc.	Sault Ste. Marie, Ontario
Almag Aluminum Inc.	Brampton, Ontario
Atlas-Graham Industries Co. Ltd.	Winnipeg, Manitoba
Comcheq Services Ltd.	Winnipeg, Manitoba
Dofasco Inc.	Hamilton, Ontario
Dun & Bradstreet Canada Limited	Mississauga, Ontario
Fisheries Products International Limited	St. John's, Newfoundland
John Wiley & Sons Canada, Ltd.	Etobicoke, Ontario
Maksteel Inc.	Mississauga, Ontario
Valley City Manufacturing Co.	Dundas, Ontario
Webcom Limited	Scarborough, Ontario

CHAPTER 19 — SAMPLE PROFIT-SHARING PLANS

COMPANY	A & G Hodgkinson Sales Ltd.
ADDRESS	P.O. Box 21059, 320 Bayfield Street Barrie, Ontario L4M 6J1
CONTACT	Mr. A.B. Hodgkinson President Telephone (705) 726-6465 Fax (705) 726-7695
NATURE OF BUSINESS	Canadian Tire associate store
NUMBER OF EMPLOYEES	73 Full-time 94 Part-time
UNION(S)	No
TYPE OF PLAN	Combination (EPSP and DPSP). Contributions that exceed the allowable limits for the DPSP are placed in the EPSP.
ORIGINAL START DATE	1965
OBJECTIVE OF PLAN	Create employee attitude conducive to professional retailing
EMPLOYER CONTRIBUTIONS	Complex formula but essentially percentage of profits
MEMBERSHIP RULES	All employees who have completed at least 1000 hours of work (Plan A) plus all employees who have completed 1–999 hours (Plan B). Dealer and his family are not members of the plan.
ALLOCATION FORMULA	Earnings (40%), merit rating (40%) and seniority (20%) to a maximum of 10,000 hours worked for Canadian Tire (not necessarily in the Barrie store or continuous service).
PAYMENT ON TERMINATION	Not applicable — see *Vesting*
PAYMENTS	Annually
PROFIT-SHARING COMMITTEE	Four members appointed by dealer. Chaired by dealer.
VESTING	Immediate — 100%

WITHDRAWALS	Allowed for purchase of a home, to reduce a mortgage or to pay for tuition at an accredited college or university.
TRUSTEE	Life insurance company.
INVESTMENT OF TRUST	DPSP — Canadian Tire class A shares EPSP — interest paid at prime rate
EMPLOYEE INVESTMENT CHOICES	None

COMPANY	Algoma Steel Inc.
ADDRESS	P.O. Box 1400 Sault Ste. Marie, Ontario P6A 5P2
CONTACT	Bernie Bos Supervisor, Wage and Salary Evaluation Telephone (705) 945-2296 Fax (705) 945-2348
NATURE OF BUSINESS	Primary manufacturing of steel
NUMBER OF EMPLOYEES	5400 Full-time
UNION(S)	U.S.W.A. Local 2251 and U.S.W.A. Local No. unknown. Total of 5300 employees in the two locals.
TYPE OF PLAN	Cash
ORIGINAL START DATE	1995
OBJECTIVE OF PLAN	To distribute a part of the company's profits to the employee-owners.
EMPLOYER CONTRIBUTIONS	Two per cent on first $40 million annual income from operations; 4% on income over $40 million to $100 million; 6% on income in excess of $100 million
MEMBERSHIP RULES	All full-time employees

ALLOCATION FORMULA	Paid on the basis of hours, i.e., employee's as a share of total hours.
PAYMENT ON TERMINATION	The share of employees who quit or retire is to be negotiated. Paid leaves of absence receive full payment. All terminations do not receive a payment. Temporary layoffs and unpaid leaves receive no payment for the period of the layoff/leave.
PAYMENTS	Annually within 90 days of the close of the fiscal year.
PROFIT-SHARING COMMITTEE	Yes, but members have not been selected yet. The selection method for both members and the chair, the term of office, and the mandate have not been determined at time of writing.

COMPANY	Almag Aluminum Inc.
ADDRESS	22 Finley Road. Bramption, Ontario L6T 1A9
CONTACT	Bob Peacock President Telephone (905) 457-9000 Fax (905) 457-9006
NATURE OF BUSINESS	Manufacture of aluminum extrusions
NUMBER OF EMPLOYEES	50 Full-time
UNION(S)	No
TYPE OF PLAN	Combination (cash and group RRSP)
ORIGINAL START DATE	1974
OBJECTIVE OF PLAN	Retirement savings (originally was 100% RRSP)

EMPLOYER CONTRIBUTIONS	Discretionary such that the group RRSP is 3–9% of earnings. Cash portion has varied from 6–17% of earnings over the past 10 years.
MEMBERSHIP RULES	All employees with at least one calendar year of service at Dec. 31.
ALLOCATION FORMULA	Pro-rated to T4 earnings
PAYMENT ON TERMINATION	All terminations receive same percentage of pay as if they were still employed.
PAYMENTS	Annually, after year-end audit
PROFIT-SHARING COMMITTEE	No

COMPANY	Atlas-Graham Industries Co. Ltd.
ADDRESS	P.O. Box 1978 Winnipeg, Manitoba R3C 3R3
CONTACT	J.A. (Joe) Graham President Telephone (204) 775-4451 Fax (204) 775-6148
NATURE OF BUSINESS	Manufacturer of brushes and mops
NUMBER OF EMPLOYEES	42 Full-time 2 Part-time 20 Temporary
UNION(S)	No
TYPE OF PLAN	Cash
ORIGINAL START DATE	1985
OBJECTIVE OF PLAN	Share profits with employees to enhance team-building efforts.
EMPLOYER CONTRIBUTIONS	Deduct 2% of sales from net profit before taxes; 30% of the result is contributed to profit-sharing pool.

MEMBERSHIP RULES	All employees from date of employment
ALLOCATION FORMULA	Equal distribution
PAYMENT ON TERMINATION	The shares of employees who are fired, quit or take leaves of absence remain in the pool. Retirees and terminations for lack of work receive a pro-rata share. Temporary lay-offs receive a full share.
PAYMENTS	Annually
PROFIT-SHARING COMMITTEE	Six members appointed by management for indefinite term. Chaired by president.

COMPANY	Comcheq Services Ltd.
ADDRESS	125 Garry Street Winnipeg, Manitoba R2C 3P2
CONTACT	Angela J. Carfrae Assistant Corporate Controller Telephone (204) 957-3822 Fax (204) 957-0855
NATURE OF BUSINESS	Computerized payroll industry
NUMBER OF EMPLOYEES	385 Full-time 176 Part-time
UNION(S)	No
TYPE OF PLAN	Combination (cash and deferred). Employee can take up to one half in cash.
ORIGINAL START DATE	1980
OBJECTIVE OF PLAN	To recognize employees' contibutions to Comcheq's success; provide retirement income for employees.
EMPLOYER CONTRIBUTIONS	Five per cent of pretax profit

MEMBERSHIP RULES	All employees other than VPs and those of higher rank; employees must work one full fiscal year (Nov. 1–Oct. 31) before being eligible.
ALLOCATION FORMULA	Based on actual number of regular hours worked (overtime not included).
PAYMENT ON TERMINATION	Quits, retirees and terminations do not receive an allocation. Employees on leaves of absence or temporary lay-off receive a pro-rata share based on hours worked.
PAYMENTS	Annually — within 120 days of year-end.
EXCESS CONTRIBUTIONS	Employee must take amounts in excess of allowable contributions to the DPSP in cash.
PROFIT-SHARING COMMITTEE	Seven members (one per region) appointed for three-year terms plus one administrator (chair) and two advisers. Employees are nominated by their peers and one name is pulled out of a hat.
VESTING	Immediate
WITHDRAWALS	Not allowed
TRUSTEE	President and CFO of Comcheq plus one independent member appointed by the executive committee of Comcheq.
INVESTMENT OF TRUST	"To get the highest return possible with little risk, at least 60% of the fund is invested in guaranteed investments." Managed by a fund manager.
EMPLOYER INVESTMENT CHOICES	None

COMPANY	Dofasco Inc.
ADDRESS	1330 Burlington Street East Hamilton, Ontario L8N 3J5
CONTACT	Paul McLenachan Manager, Pension and Benefits Telephone (905) 544-3761 Fax (905) 548-4265
NATURE OF BUSINESS	Steel manufacturer
NUMBER OF EMPLOYEES	7000 Full-time
UNION(S)	No
TYPE OF PLAN	Combination (registered pension plan, a DPSP and a group RRSP with a cash option re non-RPP amounts).
ORIGINAL START DATE	RPP—1938, DPSP—1966, group RRSP — 1989
OBJECTIVE OF PLAN	To provide retirement income; to reward employees based on company profitability.
EMPLOYER CONTRIBUTIONS	14% of pre-tax profits from manufacturing/operations.
MEMBERSHIP RULES	All full-time employees after a two-year waiting period.
ALLOCATION FORMULA	Shared equally
PAYMENT ON TERMINATION	Quits, terminations and leaves of absence do not receive a share of current year. Retirees and temporary lay-offs receive a pro-rata share based on hours worked.
PAYMENTS	Annually
EXCESS CONTRIBUTIONS	Can be taken in cash or put into group RRSP.
PROFIT-SHARING COMMITTEE	Ten appointed by management; 9 elected by employees for three-year terms. Chair is usually the CEO.

VESTING	Immediate
WITHDRAWALS	Permitted — no reason necessary. Do not usually allow more than three withdrawals per year.
TRUSTEE	Trustees appointed by management.
INVESTMENT OF TRUST	"To achieve reasonable marketability and to avoid undue illequidity through investments in short-to-medium-term debt securities. To achieve the best rate of return which is compatible with the primary objective stated above." Professional investment managers are used.
EMPLOYEE INVESTMENT CHOICES	None

∼

COMPANY	Dun & Bradstreet Canada Limited
ADDRESS	5770 Hurontario Street Mississauga, Ontario L5R 3G5
CONTACT	Helen Graham Pension and Benefits Specialist Telephone (905) 544-6335 Fax (905) 548-4265
NATURE OF BUSINESS	Business information services
NUMBER OF EMPLOYEES	465 Full-time 30 Part-time
UNION(S)	No
TYPE OF PLAN	Combination (group RRSP with a DPSP for company matches and an EPSP for RSP/DPSP excess).
ORIGINAL START DATE	1984

CHAPTER 19 — SAMPLE PROFIT-SHARING PLANS

OBJECTIVE OF PLAN	Long-term retirement savings; enhancement to pension portfolio, which contained a moderate, defined benefit, noncontributory pension plan.
EMPLOYER CONTRIBUTIONS	RRSP-employee contributions only. DPSP employer contributions of 25¢ for each dollar put in the RRSP up to 4 per cent of earnings.
MEMBERSHIP RULES	All employees with one full year of service (part-time 20+ hours and full-time)
ALLOCATION FORMULA	Based on length of service and employee contributions to RRSP
PAYMENT ON TERMINATION	All cases can cash out or transfer to other vehicles such as an RRSP. Vested funds only.
PAYMENTS	Annually
EXCESS CONTRIBUTIONS	Placed in EPSP
PROFIT-SHARING COMMITTEE	Four members of senior management. Chaired by human resources representative.
VESTING	Full vesting after two years of plan membership
WITHDRAWALS	Permitted for RRSP anytime but membership is suspended for six months. For DPSP, allowed for "financial necessity only." This is reviewed by the committee. Withdrawals must be made from the EPSP/RRSP before any are made from the deferred plan.
TRUSTEE	Trust company
INVESTMENT OF TRUST	Balanced fund; administered by profit-sharing committee.
EMPLOYEE INVESTMENT CHOICES	None

COMPANY	Fisheries Products International Limited
ADDRESS	P.O. Box 550, 70 O'Leary Avenue St. John's, Newfoundland A1C 5L1
CONTACT	Kevin J. Coombs, Vice-President, Industrial Relations Telephone (709) 570-0000 Fax (709) 570-0209
NATURE OF BUSINESS	Food processing and marketing
NUMBER OF EMPLOYEES	2,500 Full- and Part-time
UNION(S)	Fishermen, Food and Allied Workers Union
TYPE OF PLAN	Combination (cash and pension plan). First 75% of fund is paid in cash — remainder is allocated to benefit improvement; usually pension.
ORIGINAL START DATE	1986
OBJECTIVE OF PLAN	Allow employees to share in the success of the company.
EMPLOYER CONTRIBUTIONS	10% of pre-tax income
MEMBERSHIP RULES	All employees except senior management; service requirement of 30 days worked.
ALLOCATION FORMULA	Pro-rated to employee's gross earnings
PAYMENT ON TERMINATION	Retirees, all leaves of absence and temporary lay-offs are treated the same as regular employees. Quits and all terminations are not included in the plan.
PAYMENTS	Annually, just prior to Christmas
EXCESS CONTRIBUTIONS	Not an issue
PROFIT-SHARING COMMITTEE	No
VESTING	According to pension plan rules
WITHDRAWALS	Not applicable

TRUSTEE	Trust company
INVESTMENT OF TRUST	Pension committee
EMPLOYEE INVESTMENT CHOICES	None

∼

COMPANY	Maksteel Inc.
ADDRESS	7615 Torbram Road Mississauga, Ontario L4T 4A8
CONTACT	R.E. Rollwagen Vice President Telephone (905) 673-4905 Fax (905) 678-6755
NATURE OF BUSINESS	Steel service centre
NUMBER OF EMPLOYEES	250 Full-time
UNION(S)	No
TYPE OF PLAN	Cash
ORIGINAL START DATE	1970
OBJECTIVE OF PLAN	To involve employees in generating profit
EMPLOYER CONTRIBUTIONS	Specified percentage of net income before taxes. Net is calculated after a specified return to the shareholder.
MEMBERSHIP RULES	All employees with one year of service.
ALLOCATION FORMULA	Equal distribution
PAYMENT ON TERMINATION	Employees who are fired, quit or are temporarily laid off do not receive a payment. Retirees and unpaid leaves receive a pro-rata share based on time worked.

PAYMENTS	Annually on June 15.
PROFIT-SHARING COMMITTEE	No

~

COMPANY	Valley City Manufacturing Company
ADDRESS	64 Hatt Street Dundas, Ontario L9H 2G3
CONTACT	Robert Crockford President Telephone (905) 628-2253 Fax (905) 628-4470
NATURE OF BUSINESS	Architectural woodwork, cabinetry
NUMBER OF EMPLOYEES	105 Full-time
UNION(S)	Local 1057 Carpenters and Joiners — 70 members
TYPE OF PLAN	Combination (cash and DPSP). Employee has option to take cash or defer.
ORIGINAL START DATE	1964
OBJECTIVE OF PLAN	To promote harmonious working environment and reward success.
EMPLOYER CONTRIBUTIONS	27% of before-tax profits
MEMBERSHIP RULES	All employees with one year of service.
ALLOCATION FORMULA	Pro-rate on the basis of regular earnings; overtime earnings are excluded.
PAYMENT ON TERMINATION	All terminations receive a share based on their regular earnings during the period worked.
PAYMENTS	Annually.
PROFIT-SHARING COMMITTEE	Four members appointed for indefinite terms; chaired by president.

VESTING	Immediate
WITHDRAWALS	Allowed with no restrictions
TRUSTEE	Trusteed by committee members
INVESTMENT OF TRUST	Balanced investments under professional management
EMPLOYEE INVESTMENT CHOICES	None

~

COMPANY	Webcom Limited
ADDRESS	3480 Pharmacy Avenue Toronto, Ontario M1W 3G3
CONTACT	Warren D. Wilkins President Telephone (416) 496-1000 Fax (416) 496-1537
NATURE OF BUSINESS	Manufacturer of books, manuals, etc.
NUMBER OF EMPLOYEES	185 Full-time 10 Part-time
UNION(S)	No
TYPE OF PLAN	Cash
ORIGINAL START DATE	1983
OBJECTIVE OF PLAN	To provide a more rewarding worklife through increased employee involvement
EMPLOYER CONTRIBUTIONS	Specified percentages of net profits after fair return on owner's equity.
MEMBERSHIP RULES	All employees, after a waiting period of six months.
ALLOCATION FORMULA	Pro-rated to regular earnings.

PAYMENT ON TERMINATION	All terminations except those that are performance-related receive a share based on their earnings during the period worked.
PAYMENTS	Annually
PROFIT-SHARING COMMITTEE	Six members elected by fellow employees. Two-year terms chaired by president.

~

COMPANY	John Wiley & Sons Canada, Ltd.
ADDRESS	22 Worcester Road Etobicoke, Ontario M9W 1L1
CONTACT	Berni Galway Manager, Human Resources Telephone (416) 236-4433 Fax (416) 236-4447
NATURE OF BUSINESS	Publishing
NUMBER OF EMPLOYESS	70 Full-time 2 Part-time
UNION(S)	No
TYPE OF PLAN	Cash
ORIGINAL START DATE	1990
OBJECTIVE OF PLAN	To create a sense of ownership for all staff.
EMPLOYER CONTRIBUTIONS	A minimum level of operating profit is budgeted each year. A percentage of any profit in excess of this target is contributed to the pool, up to a maximum of 10% of salary for the positions eligible
MEMBERSHIP RULES	All management/sales representatives are on separate bonus plans. This plan is for non-bonusable staff: customer service, warehouse, etc. Waiting period is a probationary period from three to six months.

CHAPTER 19 — SAMPLE PROFIT-SHARING PLANS

ALLOCATION FORMULA — Pro-rate to annual base salary as of April 30 (end of fiscal year). Maximum payment is 10% of salary; minimum is 1%.

PAYMENT ON TERMINATION — Retirees and unpaid leaves are pro-rated for time worked; paid leaves and maternity leaves receive full year payment; quits must be on staff on date of payout; all others receive no payment.

PAYMENTS — Annually, on or before June 15. Fiscal year-end is April 30.

PROFIT-SHARING COMMITTEE — Executive committee: president, v.p. finance, manager human resources.

APPENDIX 1
ADDITIONAL SOURCES OF INFORMATION

The Trust Companies Association of Canada
50 rue O'Connor Street
Suite 720
Ottawa, Ontario
K1P 6L2
Telephone (613) 563-3205
Fax (613) 235-3111

Canadian Association of Suggestion Systems (CASS)
P.O. Box 55197
Fairview Mall Postal Outlet
1800 Sheppard Avenue East
North York, Ontario
M2J 5B9
Telephone (416) 490-0731

Profit Sharing Council of America
10 S. Riverside Plaza, Suite 1460
Chicago, Illinois
U.S.A. 60606-3802
Telephone (312) 441-8550
Fax (312) 441-8559

Employee Involvement Association
(formerly the National Association of Suggestion Systems)
1735 North Lynn Street
Suite 950
Arlington, Virginia 22209-2022
U.S.A.
Telephone (703) 524-3424
Fax (703) 524-2303

Bureau of Labour Information
Human Resources Development Canada (Labour Programs)
Ottawa, Ontario
Telephone (819) 953-0123 OR 1-800-567-6866
Fax (819) 953-9582

Human Resources Professionals Association of Ontario
2 Bloor Street W., Suite 1902
Toronto, Ontario
M4W 3E2
Telephone (416) 923-2324 - 1-800-387-1311
Fax (416) 923-7264

Human Resources Management Association of Greater Victoria
Box 7000
556 Boleskin Road
Victoria, B.C. V8W 2R1
Telephone (604) 361-4819
Fax (604) 361-4819

Human Resources Association of Nova Scotia
P.O. Box 592
Halifax, N.S.
B3J 2R7
Telephone (902) 860-0877
Fax (902) 860-1240

Association Des Professionels en Ressources Humaines Du Quebec
1253, avenue McGill College
Bureau 820
Montreal, Quebec
H3B 2Y5
Telephone (514) 879-1636
Fax (514) 879-1722

B.C. Human Resources Management Association
704-1130 Pender St.
Vancouver, B.C.
V6E 4A4
Telephone (604) 684-7228
Fax (604) 684-3225

Saskatchewan Council of Human Resource Associations
Box 1520
Saskatoon, Sask.
S7K 3R5
Telephone (306) 655-2449

Human Resources Institute of Alberta
Box 59
Red Deer, Alberta
T4N 5E7
Telephone 1-800-668-6125
Fax (403) 244-5060

Human Resources Management Association of Manitoba
385 St. Mary Avenue
Winnipeg, Manitoba
R3C 0N1
Telephone (204) 943-2836
Fax (204) 943-1109

Canadian Compensation Association
P.O. Box 294
Kleinberg, Ontario
L0J 1C0
Telephone (905) 893-1689
Fax (905) 893-2392

APPENDIX

2

FORMULAS FOR CHAPTER 8

This Appendix contains each of the spreadsheets included in Chapter 8. They are identical in form and layout except that, instead of the results of calculations, they show the actual formulas that are used. Where text is input in an example in Chapter 8 it is simply duplicated in this Appendix.

For example, in Option A, Cell B11 contains the name of the president, William Dixon. This shows as text in both Chapter 8 and this Appendix. In contrast, Cell B7 shows as $180,000 in Chapter 8 while here it shows as the formula: +B6*B5/100.

Since the formulas are written in Quattro Pro 5.0 for DOS, you should be aware of the various conventions that the program uses in case you wish to adjust the formulas. They are:

1. Text input should start with an '.

2. Formulas should start with a plus sign, +, otherwise they will show as text.

3. A dollar sign ($) in front of a letter or number in a formula means that if you copy that formula to another cell, that letter or number will not change. Without the $ sign Quattro

Pro adjusts the column letter or row to reflect the new location.

4. Quattro Pro uses the following arithmetic signs:

Multiplication *

Addition +

Subtraction -

Division /

The formula spreadsheets are on the following pages:

OPTION	FILE NAME	FACTOR(S) IN ALLOCATION FORMULA	PAGE CHAP 8	PAGE APPENDIX
A	A2.WQ2	Earnings	82	219
B	B2.WQ2	Job Levels	83	220
C	C2.WQ2	Seniority/Length of Service	84	221
D	D2.WQ2	Attendance	85	222
E	E2.WQ2	Employee Contributions	86	223
F	F2.WQ2	Merit Ratings	87	224
G	G2.WQ2	Equal Distribution	88	225
H	H2.WQ2	Length of Service/Earnings	89	226
I	I2.WQ2	Length of Service/Job Level	90	227
J	J1.WQ2	Merit Ratings/Earnings	91	228

Profit-Sharing Plan Worksheet
Option A2 Allocation = Base Earnings Only

	A	B	C	D	E	F	G	H	Row Number
	Fiscal Period	94-95							4
	Profits	$1,500,000							5
	% of Profits	12							6
	Fund	$180,000							7
									8
	Salary Grade/ Job Level	Name of Employee	Base Annual Salary	Other Earnings	Total Earnings	Payout	Payout as % of Base	Payout as % of Total	9
		William Dixon	$150,000	$45,000	+C11+D11	(C11/$C$26)*$B$7	(F11/C11)	+F11/E11	11
		Lynda Harris	$100,000	$22,000	+C12+D12	(C12/$C$26)*$B$7	(F12/C12)	+F12/E12	12
		Robert Campbell	$95,000	$15,000	+C13+D13	(C13/$C$26)*$B$7	(F13/C13)	+F13/E13	13
		Neil Prior	$75,000	$0	+C14+D14	(C14/$C$26)*$B$7	(F14/C14)	+F14/E14	14
		James Curtis	$72,000	$0	+C15+D15	(C15/$C$26)*$B$7	(F15/C15)	+F15/E15	15
		Jeannette Totten	$70,000	$0	+C16+D16	(C16/$C$26)*$B$7	(F16/C16)	+F16/E16	16
		Jack Kelly	$68,000	$0	+C17+D17	(C17/$C$26)*$B$7	(F17/C17)	+F17/E17	17
		Ann Walker	$65,000	$0	+C18+D18	(C18/$C$26)*$B$7	(F18/C18)	+F18/E18	18
		Nancy Dawson	$48,000	$0	+C19+D19	(C19/$C$26)*$B$7	(F19/C19)	+F19/E19	19
		Thomas Whicher	$32,000	$0	+C20+D20	(C20/$C$26)*$B$7	(F20/C20)	+F20/E20	20
		Lorrie Pierce	$32,000	$0	+C21+D21	(C21/$C$26)*$B$7	(F21/C21)	+F21/E21	21
		Susanna Jackson	$25,000	$15,500	+C22+D22	(C22/$C$26)*$B$7	(F22/C22)	+F22/E22	22
	Hourly	111	$2,664,000	$222,000	+C23+D23	(C23/$C$26)*$B$7	(F23/C23)	+F23/E23	23
	Salaried	15	$330,000	$22,500	+C24+D24	(C24/$C$26)*$B$7	(F24/C24)	+F24/E24	24
	Sales	12	$240,000	$120,000	+C25+D25	(C25/$C$26)*$B$7	(F25/C25)	+F25/E25	25
			@SUM(C11..C25)	@SUM(D11..D25)	@SUM(E11..E25)	@SUM(F11..F25)	(F26/C26)	+F26/E26	26
					+C26+D26				27

Payout Per Person for Multi-Incumbent Job Classes

Hourly =	+F23/$B23
Salaried =	+F24/$B24
Sales =	+F25/$B25

Profit-Sharing Plan Worksheet

Option B2 Allocation = Job Levels

Job Levels	%age of Fund
Top Management = A	10
Middle Management = B	25
First Level Supervisors = C	15
All others = D	50
Total	**100**

Job Levels/Equal Within Levels

	Number of Employees in Level
	Number
	3
	6
	3
	138
	150
Category of Employee	
Hourly	111
Salaried	15
Sales Reps.	12

	Fiscal Period	94-95
	Profits	$1,500,000
	% of Profits	12
	Fund	+B11*B12/100
		A	**B**

Salary Grade/Job Level	Name of Employee	Base Annual Salary	Other Earnings	Total Earnings	Payout	Payout as % of Base	Payout as % of Total
A	William Dixon	$150,000	$45,000	+C17+D17	+$E$4/100*$B$13/$F$4	(F17/C17)	+F17/E17
A	Lynda Harris	$100,000	$22,000	+C18+D18	+$E$4/100*$B$13/$F$4	(F18/C18)	+F18/E18
A	Robert Campbell	$95,000	$15,000	+C19+D19	+$E$4/100*$B$13/$F$4	(F19/C19)	+F19/E19
B	Neil Prior	$75,000	$0	+C20+D20	+$E$5/100*$B$13/$F$5	(F20/C20)	+F20/E20
B	James Curtis	$72,000	$0	+C21+D21	+$E$5/100*$B$13/$F$5	(F21/C21)	+F21/E21
B	Jeannette Totten	$70,000	$0	+C22+D22	+$E$5/100*$B$13/$F$5	(F22/C22)	+F22/E22
B	Jack Kelly	$88,000	$0	+C23+D23	+$E$5/100*$B$13/$F$5	(F23/C23)	+F23/E23
B	Ann Walker	$65,000	$0	+C24+D24	+$E$5/100*$B$13/$F$5	(F24/C24)	+F24/E24
B	Nancy Dawson	$48,000	$0	+C25+D25	+$E$5/100*$B$13/$F$5	(F25/C25)	+F25/E25
C	Thomas Whicher	$32,000	$0	+C26+D26	+$E$6/100*$B$13/$F$6	(F26/C26)	+F26/E26
C	Lorrie Pierce	$32,000	$0	+C27+D27	+$E$6/100*$B$13/$F$6	(F27/C27)	+F27/E27
C	Susanna Jackson	$25,000	$15,500	+C28+D28	+$E$6/100*$B$13/$F$6	(F28/C28)	+F28/E28
D-Hourly	111	$2,664,000	$222,000	+C29+D29	+$E$7/100*$B$13/$F$7*B29	(F29/C29)	+F29/E29
D-Salaried	15	$330,000	$22,500	+C30+D30	+$E$7/100*$B$13/$F$7*B30	(F30/C30)	+F30/E30
D-Sales Reps.	12	$240,000	$120,000	+C31+D31	+$E$7/100*$B$13/$F$7*B31	(F31/C31)	+F31/E31
		@SUM(C17..C31)	@SUM(D17..D31)	@SUM(E17..E31)	@SUM(F17..F31)	(F32/C32)	+F32/E32
				+C32+D32			

Payout Per Person for Multi-Incumbent Job Classes

Hourly =	+F29/F11
Salaried =	+F30/F12
Sales =	+F31/F13

Profit-Sharing Plan Worksheet

Option C2 Allocation = Seniority/Length of Service

Assumptions
Points per year of service = 3   With no maximum

	A	B	Row Number
Fiscal Period	94-95		4
Profits	$1,500,000		5
% of Profits	12		6
Fund	+B7*B8/100		7

Salary Grade/Job Level	Name of Employee	Base Annual Salary	Other Earnings	Total Earnings	Payout on Base	Payout as % of Base	Payout as % of Total	Years of Service	Seniority Points	Row Number
A	B	C	D	E	F	G	H	I	J	
	William Dixon	$150,000	$45,000	+C13+D13	+J13/J$28*$B$9	(F13/C13)	+F13/E13	16	48	13
	Lynda Harris	$100,000	$22,000	+C14+D14	+J14/J$28*$B$9	(F14/C14)	+F14/E14	22	66	14
	Robert Campbell	$95,000	$15,000	+C15+D15	+J15/J$28*$B$9	(F15/C15)	+F15/E15	4	12	15
	Neil Prior	$75,000	$0	+C16+D16	+J16/J$28*$B$9	(F16/C16)	+F16/E16	9	27	16
	James Curtis	$72,000	$0	+C17+D17	+J17/J$28*$B$9	(F17/C17)	+F17/E17	12	36	17
	Jeannette Totten	$70,000	$0	+C18+D18	+J18/J$28*$B$9	(F18/C18)	+F18/E18	2	6	18
	Jack Kelly	$68,000	$0	+C19+D19	+J19/J$28*$B$9	(F19/C19)	+F19/E19	7	21	19
	Ann Walker	$65,000	$0	+C20+D20	+J20/J$28*$B$9	(F20/C20)	+F20/E20	5	15	20
	Nancy Dawson	$48,000	$0	+C21+D21	+J21/J$28*$B$9	(F21/C21)	+F21/E21	1	3	21
	Thomas Whicher	$32,000	$0	+C22+D22	+J22/J$28*$B$9	(F22/C22)	+F22/E22	6	18	22
	Lorrie Pierce	$32,000	$0	+C23+D23	+J23/J$28*$B$9	(F23/C23)	+F23/E23	11	33	23
	Susanna Jackson	$25,000	$15,500	+C24+D24	+J24/J$28*$B$9	(F24/C24)	+F24/E24	6	18	24
Hourly	111	$2,664,000	$222,000	+C25+D25	+J25/J$28*$B$9	(F25/C25)	+F25/E25	4	1332	25
Salaried	15	$330,000	$22,500	+C26+D26	+J26/J$28*$B$9	(F26/C26)	+F26/E26	9	405	26
Sales Reps.	12	$240,000	$120,000	+C27+D27	+J27/J$28*$B$9	(F27/C27)	+F27/E27	8	288	27
		@SUM(C13..C27)	@SUM(D13..D27)	@SUM(E13..E27)	@SUM(F13..F27)	(F28/C28)	+F28/E28		@SUM(J13..J27)	28
				+C28+D28						29

Payout Per Person for Multi-Incumbent Job Classes

Hourly =	+F25/B25
Salaried =	+F26/B26
Sales =	+F27/B27

Profit-Sharing Plan Worksheet

Option D2 Allocation = Attendance (Regular Hours Only)

	A	B	C	D	E	F	G	H	I
Row Number									
4	**Fiscal Period**	94-95							
5	Profits	$1,500,000		**Assumptions**					
6	% of Profits	12		Average hours/year of non-management					
7	Fund	+B6*B7/100		Hourly =	2066				
8				Salaried =	2071				
9				Sales =	2075				
10									
11									
12	Salary Grade/ Job Level	Name of Employee	Base Annual Salary	Other Earnings	Total Earnings	Payout	Payout as % of Base	Payout as % of Total	Hours Worked
13		William Dixon	$150,000	$45,000	+C13+D13	+I13/I$28*$B$	(F13/C13)	+F13/E13	2061
14		Lynda Harris	$100,000	$22,000	+C14+D14	+I14/I$28*$B$	(F14/C14)	+F14/E14	2080
15		Robert Campbell	$95,000	$15,000	+C15+D15	+I15/I$28*$B$	(F15/C15)	+F15/E15	2000
16		Neil Prior	$75,000	$0	+C16+D16	+I16/I$28*$B$	(F16/C16)	+F16/E16	2045
17		James Curtis	$72,000	$0	+C17+D17	+I17/I$28*$B$	(F17/C17)	+F17/E17	2070
18		Jeannette Totten	$70,000	$0	+C18+D18	+I18/I$28*$B$	(F18/C18)	+F18/E18	2073
19		Jack Kelly	$68,000	$0	+C19+D19	+I19/I$28*$B$	(F19/C19)	+F19/E19	2080
20		Ann Walker	$65,000	$0	+C20+D20	+I20/I$28*$B$	(F20/C20)	+F20/E20	1000
21		Nancy Dawson	$48,000	$0	+C21+D21	+I21/I$28*$B$	(F21/C21)	+F21/E21	2080
22		Thomas Whicher	$32,000	$0	+C22+D22	+I22/I$28*$B$	(F22/C22)	+F22/E22	2065
23		Lorrie Pierce	$32,000	$0	+C23+D23	+I23/I$28*$B$	(F23/C23)	+F23/E23	2075
24		Susanna Jackson	$25,000	$15,500	+C24+D24	+I24/I$28*$B$	(F24/C24)	+F24/E24	2080
25	Hourly	111	$2,664,000	$222,000	+C25+D25	+I25/I$28*$B$	(F25/C25)	+F25/E25	229326
26	Salaried	15	$330,000	$22,500	+C26+D26	+I26/I$28*$B$	(F26/C26)	+F26/E26	31065
27	Sales Reps.	12	$240,000	$120,000	+C27+D27	+I27/I$28*$B$	(F27/C27)	+F27/E27	24900
28			@SUM(C13..C27)	@SUM(D13..D27)	@SUM(E13..E27)	@SUM(F13..F27)	(F28/C28)	+F28/E28	309000
29					+C28+D28				

Payout Per Person for Multi-Incumbent Job Classes

Hourly =	+F25/B25
Salaried =	+F26/B26
Sales Reps. =	+F27/B27

Profit-Sharing Plan Worksheet

Option E2 Allocation = Employee Contributions to Group R.R.S.P.

	A	B
	Fiscal Period	94-95
	Profits	$1,500,000
	% of Profits	12
	Fund	+B5*B6/100

Assumptions

Average contribution $/year of non-management employee
- Hourly = $800
- Salaried = $500
- Sales = $3,000

Employer contribution(s) for each dollar of employee contribution = 0.5

	A	B	C	D	E	F	G	H	I
Row	Salary Grade/ Job Level	Name of Employee	Base Annual Salary	Other Earnings	Total Earnings	Payout	Payout as % of Base	Payout as % of Total	Employee Contribution
14		William Dixon	$150,000	$45,000	+C14+D14	+I14*$H$10	+F14/C14	+F14/E14	$9,750
15		Lynda Harris	$100,000	$22,000	+C15+D15	+I15*$H$10	+F15/C15	+F15/E15	$7,320
16		Robert Campbell	$95,000	$15,000	+C16+D16	+I16*$H$10	+F16/C16	+F16/E16	$8,800
17		Neil Prior	$75,000	$0	+C17+D17	+I17*$H$10	+F17/C17	+F17/E17	$2,250
18		James Curtis	$72,000	$0	+C18+D18	+I18*$H$10	+F18/C18	+F18/E18	$4,320
19		Jeannette Totten	$70,000	$0	+C19+D19	+I19*$H$10	+F19/C19	+F19/E19	$0
20		Jack Kelly	$68,000	$0	+C20+D20	+I20*$H$10	+F20/C20	+F20/E20	$4,760
21		Ann Walker	$65,000	$0	+C21+D21	+I21*$H$10	+F21/C21	+F21/E21	$500
22		Nancy Dawson	$48,000	$0	+C22+D22	+I22*$H$10	+F22/C22	+F22/E22	$2,000
23		Thomas Whicher	$32,000	$0	+C23+D23	+I23*$H$10	+F23/C23	+F23/E23	$1,000
24		Lorrie Pierce	$32,000	$0	+C24+D24	+I24*$H$10	+F24/C24	+F24/E24	$1,500
25		Susanna Jackson	$25,000	$15,500	+C25+D25	+I25*$H$10	+F25/C25	+F25/E25	$3,000
26	Hourly	111	$2,664,000	$222,000	+C26+D26	+I26*$H$10	+F26/C26	+F26/E26	$88,800
27	Salaried	15	$330,000	$22,500	+C27+D27	+I27*$H$10	+F27/C27	+F27/E27	$7,500
28	Sales Reps.	12	$240,000	$120,000	+C28+D28	+I28*$H$10	+F28/C28	+F28/E28	$36,000
29			@SUM(C14..C28)	@SUM(D14..D28)	@SUM(E14..E28)	+I29*$H$10	+F29/C29	+F29/E29	$177,500
30					+C29+D29				

Payout Per Person for Multi-Incumbent Job Classes

Hourly =	+F26/B26	
Salaried =	+F27/B27	
Sales Reps. =	+F28/B28	

Profit-Sharing Plan Worksheet

Option F2 Allocation = Merit Ratings

Assumptions

Merit rating system provides points as follows:

Outstanding = A =	15
Good = B =	10
Meets job requirements = C =	5
Needs improvement = D =	2
Completely unsatisfactory = E =	0

Fiscal Period	94-95
Profits	$1,500,000
% of Profits	12
Fund	+B9*B10/100

Salary Grade/Job Level	Name of Employee	Base Annual Salary	Other Earnings	Total Earnings	Payout	Payout as % of Base	Payout as % of Total	Merit Rating	Merit Points	Row Number
	William Dixon	$150,000	$45,000	+C15+D15	+J15/$J$30*$B$11	+F15/C15	+F15/E15	B	10	15
	Lynda Harris	$100,000	$22,000	+C16+D16	+J16/$J$30*$B$11	+F16/C16	+F16/E16	A	15	16
	Robert Campbell	$95,000	$15,000	+C17+D17	+J17/$J$30*$B$11	+F17/C17	+F17/E17	C	5	17
	Neil Prior	$75,000	$0	+C18+D18	+J18/$J$30*$B$11	+F18/C18	+F18/E18	C	5	18
	James Curtis	$72,000	$0	+C19+D19	+J19/$J$30*$B$11	+F19/C19	+F19/E19	C	5	19
	Jeannette Totten	$70,000	$0	+C20+D20	+J20/$J$30*$B$11	+F20/C20	+F20/E20	D	2	20
	Jack Kelly	$68,000	$0	+C21+D21	+J21/$J$30*$B$11	+F21/C21	+F21/E21	B	10	21
	Ann Walker	$65,000	$0	+C22+D22	+J22/$J$30*$B$11	+F22/C22	+F22/E22	C	5	22
	Nancy Dawson	$48,000	$0	+C23+D23	+J23/$J$30*$B$11	+F23/C23	+F23/E23	A	15	23
	Thomas Whicher	$32,000	$0	+C24+D24	+J24/$J$30*$B$11	+F24/C24	+F24/E24	B	10	24
	Lorrie Pierce	$32,000	$0	+C25+D25	+J25/$J$30*$B$11	+F25/C25	+F25/E25	A	15	25
	Susanna Jackson	$25,000	$15,500	+C26+D26	+J26/$J$30*$B$11	+F26/C26	+F26/E26	B	10	26
Hourly	111	$2,664,000	$222,000	+C27+D27	+J27/$J$30*$B$11	+F27/C27	+F27/E27	C	555	27
Salaried	15	$330,000	$22,500	+C28+D28	+J28/$J$30*$B$11	+F28/C28	+F28/E28	C	75	28
Sales Reps.	12	$240,000	$120,000	+C29+D29	+J29/$J$30*$B$11	+F29/C29	+F29/E29	C	60	29
		@SUM(C15.C29)	@SUM(D15.D29)	@SUM(E15.E29)	@SUM(F15.F29)	+F30/C30	+F30/E30		797	30
				+C30+D30		+F30/C30				31
									5.3	

Payout Per Person for Multi-incumbent Job Classes

Hourly =	+F27/B27
Salaried =	+F28/B28
Sales Reps. =	+F29/B29

Profit-Sharing Plan Worksheet

Option G2 Allocation = Equal Distribution

	A	B	C	D	E	F	G	H	Row Number
Fiscal Period	94-95								4
Profits	$1,500,000				Number of Employees		150		5
% of Profits	12								6
Fund	+B5*B6/100								7
					Assumption				8
Salary Grade/ Job Level		Name of Employee	Base Annual Salary	Other Earnings	Total Earnings	Payout	Payout as % of Base	Payout as % of Total	9
									10
		William Dixon	$150,000	$45,000	+C11+D11	+B$7/$G$4	+F11/C11	+F11/E11	11
		Lynda Harris	$100,000	$22,000	+C12+D12	+B$7/$G$4	+F12/C12	+F12/E12	12
		Robert Campbell	$95,000	$15,000	+C13+D13	+B$7/$G$4	+F13/C13	+F13/E13	13
		Neil Prior	$75,000	$0	+C14+D14	+B$7/$G$4	+F14/C14	+F14/E14	14
		James Curtis	$72,000	$0	+C15+D15	+B$7/$G$4	+F15/C15	+F15/E15	15
		Jeannette Totten	$70,000	$0	+C16+D16	+B$7/$G$4	+F16/C16	+F16/E16	16
		Jack Kelly	$68,000	$0	+C17+D17	+B$7/$G$4	+F17/C17	+F17/E17	17
		Ann Walker	$65,000	$0	+C18+D18	+B$7/$G$4	+F18/C18	+F18/E18	18
		Nancy Dawson	$48,000	$0	+C19+D19	+B$7/$G$4	+F19/C19	+F19/E19	19
		Thomas Whicher	$32,000	$0	+C20+D20	+B$7/$G$4	+F20/C20	+F20/E20	20
		Lorrie Pierce	$32,000	$15,500	+C21+D21	+B$7/$G$4	+F21/C21	+F21/E21	21
		Susanna Jackson	$25,000	$222,000	+C22+D22	+B$7/$G$4	+F22/C22	+F22/E22	22
Hourly		111	$2,664,000		+C23+D23	+B$7/$G$4*B23	+F23/C23	+F23/E23	23
Salaried		15	$330,000	$22,500	+C24+D24	+B$7/$G$4*B24	+F24/C24	+F24/E24	24
Sales Reps.		12	$240,000	$120,000	+C25+D25	+B$7/$G$4*B25	+F25/C25	+F25/E25	25
			@SUM(C11..C25)	@SUM(D11..D25)	@SUM(E11..E25)	@SUM(F11..F25)	+F26/C26	+F26/E26	26
		@COUNT(B11..B22)			+C26+D26				27
		@SUM(B23..B25)	+B27+B28						28
Payout Per Person for Multi-Incumbent Job Classes									29
Hourly =	+F23/B23								30
Salaried =	+F24/B24								31
Sales Reps. =	+F25/B25								32

Profit-Sharing Plan Worksheet

Option H2 Allocation = Length of Service/Base Earnings

	A	B	C	D	E	F	G	H	I	Row Number
Fiscal Period	94-95									4
Profits	$1,500,000									5
% of Profits	12									6
Fund	+B5*B6/100									7
					Assumptions					8
					Points per year of service =		2			9
					Points per $1,000 of base =		1			10
Salary Grade/ Job Level	Name of Employee	Base Annual Salary	Other Earnings	Total Earnings	Average years of service =	Hourly 5	Salaried 3	Sales Reps 1		
					Payout	Payout as % of Base	Payout as % of Total	Years of Service		
	William Dixon	$150,000	$45,000	+C11+D11	((C11/1000*$G$5)+($G$4*111))/F$28*$B$7	(F11/C11)	+F11/E11	5	11
	Lynda Harris	$100,000	$22,000	+C12+D12	((C12/1000*$G$5)+($G$4*112))/F$28*$B$7	(F12/C12)	+F12/E12	3	12
	Robert Campbell	$95,000	$15,000	+C13+D13	((C13/1000*$G$5)+($G$4*113))/F$28*$B$7	(F13/C13)	+F13/E13	4	13
	Neil Prior	$75,000	$0	+C14+D14	((C14/1000*$G$5)+($G$4*114))/F$28*$B$7	(F14/C14)	+F14/E14	3	14
	James Curtis	$72,000	$0	+C15+D15	((C15/1000*$G$5)+($G$4*115))/F$28*$B$7	(F15/C15)	+F15/E15	5	15
	Jeannette Totten	$70,000	$0	+C16+D16	((C16/1000*$G$5)+($G$4*116))/F$28*$B$7	(F16/C16)	+F16/E16	2	16
	Jack Kelly	$68,000	$0	+C17+D17	((C17/1000*$G$5)+($G$4*117))/F$28*$B$7	(F17/C17)	+F17/E17	5	17
	Ann Walker	$65,000	$0	+C18+D18	((C18/1000*$G$5)+($G$4*118))/F$28*$B$7	(F18/C18)	+F18/E18	1	18
	Nancy Dawson	$48,000	$0	+C19+D19	((C19/1000*$G$5)+($G$4*119))/F$28*$B$7	(F19/C19)	+F19/E19	4	19
	Thomas Whicher	$32,000	$0	+C20+D20	((C20/1000*$G$5)+($G$4*120))/F$28*$B$7	(F20/C20)	+F20/E20	3	20
	Lorrie Pierce	$32,000	$0	+C21+D21	((C21/1000*$G$5)+($G$4*121))/F$28*$B$7	(F21/C21)	+F21/E21	5	21
	Susanna Jackson	$25,000	$15,500	+C22+D22	((C22/1000*$G$5)+($G$4*122))/F$28*$B$7	(F22/C22)	+F22/E22	1	22
Hourly	111	$2,664,000	$222,000	+C23+D23	((C23/1000*$G$5)+($G$4*123))/F$28*$B$7	(F23/C23)	+F23/E23	+F7*B23	23
Salaried	15	$330,000	$22,500	+C24+D24	((C24/1000*$G$5)+($G$4*124))/F$28*$B$7	(F24/C24)	+F24/E24	+G7*B24	24
Sales Reps.	12	$240,000	$120,000	+C25+D25	((C25/1000*$G$5)+($G$4*125))/F$28*$B$7	(F25/C25)	+F25/E25	+H7*B25	25
		@SUM(C11..C25)	@SUM(D11..D25)	@SUM(E11..E25)	@SUM(F11..F25)	(F26/C26)	+F26/E26	653		26
	Salary Points =	+C26/1000*$G$5		+C26+D26			Total Service Points	+I26*$G$4		27
		Total Salary and Service Points =		+C27+I27						28
	Payout Per Person for Multi-Incumbent Job Classes									29
	Hourly =	+F23/$B23								30
	Salaried =	+F24/$B24								31
	Sales Reps. =	+F25/$B25								32

Profit Sharing-Plan Worksheet

Fiscal Period	94-95
Profits	$1,500,000
% of Profits	12
Fund	$180,000

Option I2 Allocation = Length of Service/Job Level

Assumptions

Job Levels	%age of Fund	Number of Employees
Top Management = A	10	3
Middle Management = B	25	6
Supervisors/Professionals = C	15	3
All Others = D =	50	138

	Within Category D	Average Years
Hourly	111	5
Salaried	15	3
Sales	12	1

Points/year of service 6

	A	B	C	D	E	F	G	H	I	J
	Salary Grade/ Job Level	Name of Employee	Base Annual Salary	Other Earnings	Total Earnings	Payout	Payout as % of Base	Payout as % of Total	Years of Service	Service Points
	A	William Dixon	$150,000	$45,000	+C18-D18	+$G$5*$B$7/100*(J18/@SUM(J$18..J$20))	(F18/C18)	+F18/E18	5	+$G$13*I18
	A	Lynda Harris	$100,000	$22,000	+C19-D19	+$G$5*$B$7/100*(J19/@SUM(J$18..J$20))	(F19/C19)	+F19/E19	3	+$G$13*I19
	A	Robert Campbell	$95,000	$15,000	+C20-D20	+$G$5*$B$7/100*(J20/@SUM(J$18..J$20))	(F20/C20)	+F20/E20	4	+$G$13*I20
	B	Neil Prior	$75,000	$0	+C21+D21	+$G$6*$B$7/100*(J21/@SUM(J$21..J$26))	(F21/C21)	+F21/E21	3	+$G$13*I21
	B	James Curtis	$72,000	$0	+C22+D22	+$G$6*$B$7/100*(J22/@SUM(J$21..J$26))	(F22/C22)	+F22/E22	5	+$G$13*I22
	B	Jeannette Totten	$70,000	$0	+C23+D23	+$G$6*$B$7/100*(J23/@SUM(J$21..J$26))	(F23/C23)	+F23/E23	5	+$G$13*I23
	B	Jack Kelly	$68,000	$0	+C24+D24	+$G$6*$B$7/100*(J24/@SUM(J$21..J$26))	(F24/C24)	+F24/E24	5	+$G$13*I24
	B	Ann Walker	$65,000	$0	+C25+D25	+$G$6*$B$7/100*(J25/@SUM(J$21..J$26))	(F25/C25)	+F25/E25	1	+$G$13*I25
	B	Nancy Dawson	$48,000	$0	+C26+D26	+$G$6*$B$7/100*(J26/@SUM(J$21..J$26))	(F26/C26)	+F26/E26	4	+$G$13*I26
	C	Thomas Whicher	$32,000	$0	+C27+D27	+$G$7*$B$7/100*(J27/@SUM(J$27..J$29))	(F27/C27)	+F27/E27	5	+$G$13*I27
	C	Lorrie Pierce	$32,000	$0	+C28+D28	+$G$7*$B$7/100*(J28/@SUM(J$27..J$29))	(F28/C28)	+F28/E28	10	+$G$13*I28
	C	Susanna Jackson	$25,000	$15,500	+C29+D29	+$G$7*$B$7/100*(J29/@SUM(J$27..J$29))	(F29/C29)	+F29/E29	5	+$G$13*I29
	D - Hourly	111	$2,664,000	$222,000	+C30+D30	+$G$8*$B$7/100*(J30/@SUM(J$30..J$32))	(F30/C30)	+F30/E30	+I10*B30	+$G$13*I30
	D - Salaried	15	$330,000	$22,500	+C31+D31	+$G$8*$B$7/100*(J31/@SUM(J$30..J$32))	(F31/C31)	+F31/E31	+I11*B31	+$G$13*I31
	D - Sales Reps.	12	$240,000	$120,000	+C32+D32	+$G$8*$B$7/100*(J32/@SUM(J$30..J$32))	(F32/C32)	+F32/E32	+I12*B32	+$G$13*I32
			@SUM(C18..C32)	@SUM(D18..D32)	@SUM(E18..E32)	@SUM(F18..F32)		@SUM(F18..F32)		
					+C33-D33					

Payout Per Person for Multi-Incumbent Job Classes

Hourly =	+F30/$B30
Salaried =	+F31/$B31
Sales Reps. =	+F32/$B32

Profit Sharing-Plan Worksheet

Fiscal Period	94-95
Profits	$1,500,000
% of Profits	12
Fund	$180,000

Option J2 Allocation = Merit Ratings/Base Earnings

Assumptions

Merit rating	Points	Code
Outstanding =	15	A
Good =	10	B
Satisfactory =	5	C
Needs improvement =	2	D
Unsatisfactory =	0	E

Points per $1000 of base earnings = 2

	A	B	C	D	E	F	G	H	I	J	Row Number
	Salary Grade/ Job Level	Name of Employee	Base Annual Salary	Other Earnings	Total Earnings	Payout	Payout as % of Base	Payout as % of Total	Merit Rating	Merit Points	
											4
											5
											6
											7
											8
											9
											10
											11
											12
		William Dixon	$150,000	$45,000	+C15+D15	((C15/1000*$F$10)+J15)/$I$32*$B$7	(F15/C15)	+F15/E15	B	10	13
		Lynda Harris	$100,000	$22,000	+C16+D16	((C16/1000*$F$10)+J16)/$I$32*$B$7	(F16/C16)	+F16/E16	A	15	14
		Robert Campbell	$95,000	$15,000	+C17+D17	((C17/1000*$F$10)+J17)/$I$32*$B$7	(F17/C17)	+F17/E17	C	5	15
		Neil Prior	$75,000	$0	+C18+D18	((C18/1000*$F$10)+J18)/$I$32*$B$7	(F18/C18)	+F18/E18	C	5	16
		James Curtis	$72,000	$0	+C19+D19	((C19/1000*$F$10)+J19)/$I$32*$B$7	(F19/C19)	+F19/E19	C	5	17
		Jeannette Totten	$70,000	$0	+C20+D20	((C20/1000*$F$10)+J20)/$I$32*$B$7	(F20/C20)	+F20/E20	D	2	18
		Jack Kelly	$68,000	$0	+C21+D21	((C21/1000*$F$10)+J21)/$I$32*$B$7	(F21/C21)	+F21/E21	B	10	19
		Ann Walker	$65,000	$0	+C22+D22	((C22/1000*$F$10)+J22)/$I$32*$B$7	(F22/C22)	+F22/E22	C	5	20
		Nancy Dawson	$48,000	$0	+C23+D23	((C23/1000*$F$10)+J23)/$I$32*$B$7	(F23/C23)	+F23/E23	A	15	21
		Thomas Whicher	$32,000	$0	+C24+D24	((C24/1000*$F$10)+J24)/$I$32*$B$7	(F24/C24)	+F24/E24	B	10	22
		Lorrie Pierce	$32,000	$0	+C25+D25	((C25/1000*$F$10)+J25)/$I$32*$B$7	(F25/C25)	+F25/E25	A	15	23
		Susanna Jackson	$25,000	$15,500	+C26+D26	((C26/1000*$F$10)+J26)/$I$32*$B$7	(F26/C26)	+F26/E26	B	10	24
	Hourly	111	$2,664,000	$222,000	+C27+D27	((C27/1000*$F$10)+J27)/$I$32*$B$7	(F27/C27)	+F27/E27	C	555	25
	Salaried	15	$330,000	$22,500	+C28+D28	((C28/1000*$F$10)+J28)/$I$32*$B$7	(F28/C28)	+F28/E28	B	150	26
	Sales Reps.	12	$240,000	$120,000	+C29+D29	((C29/1000*$F$10)+J29)/$I$32*$B$7	(F29/C29)	+F29/E29	C	60	27
			@SUM(C15..C29)	@SUM(D15..D29)	@SUM(E15..E29)	@SUM(F15..F29)	(F30/C30)	+F30/E30		@SUM(J15..J29)	28
					+C30+D30						29

Payout Per Person for Multi-Incumbent Job Classes			Total Earnings Points =	+C30/1000*$F$10		Earnings Plus Merit Points	=	+$J$30÷$E$32		30
Hourly =	+F27/$B$27									31
Salaried =	+F28/$B$28									32
Sales =	+F29/$B$29									33

Note

The formula for Cell J15 is as follows:

@IF($I15=$G$5,$F$5,@IF($I15=$G$6,$F$6,@IF($I15=$G$7,$F$7,@IF($I15=$G$8,$F$8,@IF($I15=$G$9,$F$9,0)))))

The rest of the column entries are copies of this formula adjusted for Row number.

INDEX

A&G Hodgkinson Sales Ltd., 197, 198
Administration, 27.
  See also Permanent committee
Algoma Steel Inc., 55, 198, 199
Allocation formula, 57
Allocation of funds, 57-71
  assumptions, 58
  attendance, 63, 64, 85, 222
  choosing appropriate formula, 70, 71
  combination formulas, 67
  earnings, 58-60, 82, 219
  employee contributions, 64, 65, 86, 223
  equal distribution, 66, 88, 225
  factors considered, 57, 58, 67
  job levels, 60, 61, 83, 220
  length of service/earnings, 67, 68, 89, 226
  length of service/job level, 68, 69, 90, 227
  merit ratings, 65, 66, 87, 224
  merit ratings/earnings, 69, 91, 228
  seniority/length of service, 61, 62, 84, 221
  See also Testing allocation formula
Almag Aluminum Inc., 199, 200
Amount, 44-46
Announcement, 38, 39
Atlas-Graham Industries Co. Ltd., 200, 201

Briggs Collieries, 2
Broad-based plans, 7, 8

Canadian labour law, 177, 178
Cash plans, 8-10
Checklists, 131-146
Coaching employees, 153
Combination plans, 9, 12
Comcheq Services Ltd., 201, 202
Committee, *see* Design committee, Permanent committee
Communications:
  after implementation, 114-118
  documentation, 120, 121
  during design phase, 113, 114
  what to communicate, 118-120
Compensation policy, 47, 48
Conditions needed for profit sharing, 21-29
Confidentiality, 80, 81, 118
Contribution formula, 45
Copy-cat plans, 40
Creative thinking, 150

Declared formulas, 43, 44
Deferred profit-sharing plans (DPSPs):
  administration of investments, 101, 102
  advantages/disadvantages, 10, 11
  CCPC owners excluded, 55
  contributions, 103
  employee choice of investments, 100, 101
  employee contributions, 103
  establishment of trust fund, 104
  excess contributions, 103, 104
  forfeitures, 95, 96
  investment in own company's shares, 99
  investment policy, 98-101
  investment training, 149, 150
  leaving plan, 102, 103
  membership, 94
  overview, 8
  qualified investments, 99
  registration with Revenue Canada, 104
  setting up plan, 104
  vesting, 94, 95
  withdrawals, 96-98
Design committee:
  assumption, 40
  confidentiality of test results, 80, 81
  establishment, 36-38
  first meeting, 40-42, 135-137
  ideal size, 36
  meetings with employees, 138
  multiple locations, 38
  second meeting, 141-143
  third meeting, 145
  who are members, 36, 37
  See also Permanent committee
Design phase, 33-42, 113, 114
Discretionary formula, 43, 44
Diskette, 58, 74, 75
Diversity management, 154
Documentation, 120, 121
Dofasco Inc., 51, 203, 204
Dun & Bradstreet Canada Limited, 204, 205

Early Canadian plans, 2, 3
Eastman Kodak, 1
Eligibility period, 51, 52
Employee attitude/opinion surveys, 22, 23
Employee involvement:
  designing plan, 36-38

profit-sharing committee, *see* Design committee, Permanent committee
suggestions, *see* Employee suggestion systems
Employee profit-sharing plans (EPSPs):
  advantages/disadvantages, 12
  employee choice of investment, 110
  employee contributions, 111
  employer contributions, 110
  establishment of trust fund, 111
  forfeitures, 108
  investment training, 149, 150
  investments, 109, 110
  leaving plan, 111
  membership, 106
  overview, 9
  registration with Revenue Canada, 111
  setting up plan, 111
  vesting, 106, 107
  withdrawals, 109
Employee suggestion systems, 27, 28, 155-166
  awards, 164, 165
  changes in equipment/processes, 166
  disadvantages, 157
  evaluation of suggestions, 162-164
  membership eligibility, 159, 160
  protection of suggestions, 165
  submitting suggestions, 160-162
  suggestion form, 161
  supervisory training, 116
  training concerns, 116, 150, 151
  when to establish system, 157, 158
  why needed, 155-157
Employer contributions, 43-48
English as a Second Language (ESL), 150
Entitlement, *see* Membership
Equity theory, 170
Examples, *see* Testing allocation formula
Expectancy theory, 169, 170

Financial reporting, 118-120
Financial training, 148, 149
Fisheries Products International Limited, 55, 206, 207
Floppy diskette, 58, 74, 75
Foreign operations, 187, 191-193
Forfeitures:
  deferred profit-sharing plans, 95, 96
  employee profit-sharing plans, 108
Formula, 43, 44
401(k) plans, 188, 189
Frequency/timing of payments, 46, 47

Gallatin, Albert, 1
Graduated vesting, 106, 107

Immediate vesting, 94, 95, 106
Implementation, 146
Improshare, 172, 173
Individual bonus plans, 172, 174
Information, sources of, 23, 29-32, 213-216
Internal equity, 24
Investment training, 149, 150

Investments:
  deferred profit-sharing plans, 98-101
  employee profit-sharing plans, 109, 110

John Wiley & Sons Ltd., 210, 211

Language training, 150
Lead hands, 152
Leclaire, Jean, 1

Major issues/components, 57, 73
Maksteel Inc., 207, 208
Management commitment, 27-29
Membership, 49-55
  category of employment, 50, 51
  commissioned sales stuff, 53, 54
  deferred profit-sharing plans, 55, 94
  employee suggestion systems, and, 159, 160
  employee profit-sharing plans, 106
  illness, 52
  laid-off employees, 53
  length of service, 51, 52
  retroactivity, 52
  terminated employees, 52, 53
  union status, 54, 55
Merit increases, 171, 174
Mexico, 189-191
Money purchase limit, 103
Motivation, 168-170

North American Free Trade Agreement (NAFTA), 187

Objectives, 15, 34, 35

Pay-for-performance:
  defined, 170, 171
  overview of plan differences, 174
  types of plans, 171, 172
Performance management, 153
Performance measures, 151, 152
Permanent committee, 123-130
  chairperson, 125
  composition/size, 124, 125
  dates of appointment, 126, 127
  day-to-day administration of plan, 129
  deferred profit-sharing plans, and, 105
  duties, 128, 129
  importance, 124
  interim elections, 127
  re-election, 127
  review of plan, 129, 130
  selection/election, 125
  term of office, 126
  transition from design committee, 128
  See also Design committee
Pierce, Lorrie, 58
Pre-retirement planning, 117, 118
Procter and Gamble, 1
Productivity gain-sharing plans, 172-174
Profit pool, 25
Profit-sharing committee, *see* Design committee, Permanent committee
Profit Sharing Council of Canada, 45
Profit-sharing plans:
  day-to-day administration of, 129
  early Canadian plans, 2, 3

effectiveness, 13-19
history, 1-5
initial size of profit pool, 25
overview, 174
pay-for-performance plan, as, 172, 175
prevalence, 1-5, 19
productivity gain-sharing plans, contrasted, 173, 174
reasons for introduction, 15, 34, 35
reputation of company, and, 17, 18
samples, 195-211
types, 7-9
Profits, defined, 43
Prudent person rule, 101

Reports, *see* Studies/reports/surveys
Research and planning, 133, 134
Review of plan, 129, 130
Rucker plan, 172, 173

S.C. Johnson Wax, 47
Salary increase plans, 171, 174
Salary/wages:
   externally competitive, 23, 47, 48
   internal equity, 24
   sources of information, 23, 29-32
Sample profit-sharing plans, 195-211
Scanlon plan, 172, 173
Sears Roebuck, 1
Sources of information, 23, 29-32, 213-216
Spreadsheet examples, *see* Testing allocation formula

Studies/reports/surveys:
   Bell/Hanson study (G.B.), 2, 16
   Blinder paper, 17
   Board of Trade (Metro Toronto), 5
   CASS Survey, 164-166
   Hewitt Survey, 4, 19, 43, 46
   Long report, 4, 5, 14, 15, 19
   People, Performance and Pay (U.S.), 14-16, 156, 157
   Smith's study, 184
Suggestion plan administrator (SPA), 162
Supervisory training, 116, 152-154
Surveys, *see* Studies/reports/surveys

Teams, 153
Terminated employees, 52, 53
   deferred profit-sharing plans, 102, 103
   employee profit-sharing plans, 111
Testing allocation formula, 73-91
   assumptions, 75, 76
   communicating results to committee, 80, 81
   confidentiality of results, 80, 81
   examples (formula), 217-228
   examples (results of calculations), 82-91
   floppy diskette, 58, 74, 75
   general rules, 77-79
   spreadsheet program used, 74
   steps to follow, 79, 80
   tests of reasonableness, 79
   why important, 74

See also Allocation of funds
Testing models, 144
Theory X managers, 156
Top-hat plans, 7, 8
Training:
 funding/financial support, 154
 general (all employees), 147-152
 managerial/supervisory employees, 152-154
 tips, 154
Two year vesting, 94, 95

Unions, 177-185
 installation of plan, 181-184
 lists of companies with unions and plans, 180, 181
 membership in plan, and, 54, 55
 trends, 184

views of labour, 178, 179, 185
United States, 188, 189

Valley City Manufacturing Company, 55, 208, 209
Vesting:
 deferred profit-sharing plans, 94, 95
 employee profit-sharing plans, 106, 107

W.F. Hatheway Company, The, 2
Webcom Limited, 209, 210
Withdrawals:
 deferred profit-sharing plans, 96-98
 employee profit-sharing plans, 109